Out *of the* Darkness

Out *of the* Darkness

A Novel

❧ Book One of the Courageous Series ❧

DAVID A. JACINTO

Forefront
BOOKS
❧ 3 ❧

Out of the Darkness

Published by Forefront Books.
Distributed by Simon & Schuster.

Library of Congress Control Number: 2023912989

Print ISBN: 978-1-63763-189-8
E-book ISBN: 978-1-63763-190-4

Cover Design by Mimi Bark.
Interior Design by PerfecType, Nashville, TN

This book is dedicated to my family, past and present, but especially my wife Anne, also an immigrant, and related to a long line of Welsh coal miners.

I also want to dedicate this book to my four wonderful children and their spouses: Michael and Sandra; Paul and Elizabeth; Rachel and Jason; Daniel and Christie. And of course, to my near-perfect grandchildren—all thirteen of them: Alan, Kara, Clark, Cole, Jack, Miles, Eden, Parker, Cooper, Kalani, Elayna, Gionni, Xenaya.

Family Tree

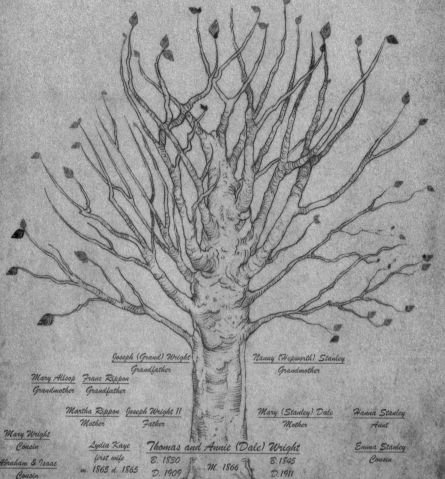

Joseph (Grand) Wright
Grandfather

Nanny (Hepworth) Stanley
Grandmother

Mary Allsop   Franc Rippon
Grandmother   Grandfather

Martha Rippon   Joseph Wright II
Mother   Father

Mary (Stanley) Dale   Hanna Stanley
Mother   Aunt

Mary Wright
Cousin

Lydia Kaye
first wife
m. 1865 d. 1865

Thomas and Annie (Dale) Wright

Emma Stanley
Cousin

Abraham & Isaac
Cousin

B. 1830
D. 1909

M. 1866

B. 1845
D. 1911

George Wright   Joseph Wright III
Brother   Brother

Edith (Wright) Hobson   Andrew Hobson
Sister   Brother-in-law

Georgie Wright II   Thomas Francis Wright
Son   Son

Thomas Hobson   Elizabeth Hobson
Nephew   Niece

Will & Emily Wright
Cousins

Zina Mae Wright   Norma Jean Wright
Granddaughter   Granddaughter

Eileen Wright
Great Granddaughter

David Jacinto
Great. Great Grandson

Introduction

✢ · ✢

IN THE EARLY NINETEENTH CENTURY, the Industrial Revolution raged across England, spawning innovation, driving a booming economy, and changing the lives of millions around the world. This cataclysmic change in world history brought manufacturing, steel mills, a web of railways, shipping ports, and coal mines unmatched anywhere in the world. England's rapidly expanding manufacturing might produced nearly all the United Kingdom's goods, 70 percent of all products in Europe, and virtually all of their coal. England became an industrial power that would dominate world trade for a century. But left in the wake were millions of desperately poor laborers enslaved by powerful aristocrats.

The turbulent transition from the antiquity of rural life to the center of industry brought with it incredible heartache and despair. Workers were massed, stacked, and crammed into industrial cities and company-owned coal mining towns without utilities, clean water, or sanitary facilities. Everything was owned by the company, including the families who made these aristocrats millions. They worked under extremely hazardous conditions, with few safety protections.

Children who did not die of infectious disease spent their nights in squalid hovels infested with lice, dysentery, and plagues of every kind.

All but the smallest usually spent their days in factories, steel mills, or a thousand feet underground in coal mines. Children were often sentenced to the death row of the assembly floor, gas kiln, or dangerous coal mines. Drawing putrid air into their lungs, they labored from the predawn darkness to well after the sun set in the west. The average life expectancy for those working in the industrial centers was twenty-two years, and those in the coal mines didn't fare much better. Few were given an opportunity to climb out of the brutality of their lives. When inevitable disaster struck, there was no compensation for the families of those who lost their lives and livelihood. Those left behind were often set adrift to survive under their own devices or thrown into workhouses where they too would eventually die.

Under these horrific conditions, England and the rest of Europe rushed reckless and unrepentant into a new era of change, a time and place that spawned heartbreaking stories by Dickens, Disraeli, and Somerset Maugham. To escape this debased human bondage, desperate workers from all over Europe flooded into America. They came in greater numbers than at any other time in history, leaving behind starvation and industrial slavery imposed by an aristocratic class. And though they brought with them differing cultures, they came with a common belief that anything was possible in America. These poor, wretched, huddled masses, hungering to breathe free, passed through that golden door in quest of a common goal—to be stakeholders in the promised land, where all people are created equal, endowed by their Creator with certain unalienable rights to life, liberty, religious freedom, and the pursuit of happiness.

꒦ · ꒦

Written on the inside of an old ship's locker I inherited from my great aunt Norma is the inscription "Martha Wright, 1810." The two-hundred-year-old trunk is filled to the brim with artifacts, archives, and journal writings of my ancestors, including a large leather-bound volume of letters, journal entries, and stories entitled "The Wrights, They Came from England."

After dragging the trunks to my attic, I sat on the dusty couch among the boxes late into the night, combing through artifacts and reading the fascinating stories of generations gone before. I was exhausted when I closed my eyes in the shadowed darkness that smelled of mildew and bygone lives. I tried to imagine casting a line to my forebears, thrown across years, continents, and generations long since passed. This was my family! I held in my hands the stories of their lives beginning more than two centuries before—stories of adversity, power, greed, struggle, overcoming, love, and family. These lives were of me, fallen leaves from the same family tree.

With that knowledge I launched a four-year investigative journey. I studied books, journals, and the web and spoke to dozens of people in and around the coal mining towns and the fabulously wealthy families of England. That investigation uncovered scattered stories of unbelievable poverty, great accomplishment, incomprehensible wealth, horrific disasters, human tragedy, the growing pains of a new nation and, of course, romance.

This novel is inspired by actual events in my ancestors' lives, who came to America not just for themselves, but for their children and the generations yet unborn. For these immigrants, desperate to unleash the great sweep of human desire, America was more than a country. It was an ideal, a miracle, "a city on the hill . . . a light unto all of the world. . . !"

❧ PART I ❧

South Yorkshire, England, 1837

Chapter One

⤜ · ⤛

SEVEN-YEAR-OLD TOMMY WRIGHT SAT IN the corner of his family's coal miner shanty. His thoughts were lost in the pages of a
book as an evening storm from the Irish Sea slipped unnoticed over
the Peak Mountains and onto the rolling hills of South Yorkshire.
Moonlight filtering through the gathering storm clouds reflected off
the open wounds of hollowed-out hillsides, a ghostly, ever-present
reminder of nature's wilderness being plundered for its mineral riches.

The shrill shriek of the Silkstone coal mine's quitting time whistle startled Tommy back into the moment. Leaving his imagination
stranded with *Robinson Crusoe* on a far-off paradise island, Tommy
turned his attention toward the only window in the tiny hovel. The
frosted windowpane splintered the soft light of the rising moon into
a rainbow of color. He drew in a shallow, shuddered breath, and with
the back of his tattered sleeve wiped the windowpane clean to peer
down the shadowed pathway toward the mine.

"Please, Papa, come home to us tonight," Tommy begged.

Today was Tommy's seventh birthday. The family had long
planned a dinner celebration on this, Tommy's first day of his becoming a man. After today he would leave the rest of the family and

his Mam's schoolroom behind to join his papa down in the pit. The Wright family had worked the collieries along this seam of coal, the largest in all of England, for more generations than anyone could remember. And like his papa and grand before him, Tommy would take his place in the family business, mining coal twelve hours a day, six days a week. Only on Sundays could he expect to see the sun for the rest of his life. His pay would go toward the Wright family's tiny, one-room shanty, blackened by coal dust on the outside and dingy grey soot from the coal-fired Higgins hearth on the inside. Without utilities or conveniences of any kind, theirs was one of dozens of hastily built company dwellings in Silkstone's ramshackle shantytown located just outside the gates of the coal mine.

Impatient, Tommy peered through the cracked windowpane at the passing stream of raggedy-clothed miners. Rain-soaked and shivering, men and boys left behind the low industrial hum of the mine, and man's war against nature, their eyes glowing white out of blackened faces. Hunched over, with knapsacks slung across their shoulders, they trudged up the muddy pathway in strung-out lines. But Papa was not among them.

Tommy knew all too well that most of the miners would have already peeled off toward the company pub before coming into view. Most miners spent their evenings at the pub, where they laid aside their troubled lives for at least a few hours, hoping to drown out their coal miner's gloom in a pint of ale. In drunken camaraderie, they sat in front of a warm fire in the great hearth, telling stories of better times and better places, all the while sinking further in debt on company credit.

On more payday nights than he cared to remember, Tommy and his three younger siblings sat with his somber Mam outside the company pub waiting for Papa. Martha, the rather diminutive, twenty-seven-year-old mother of four, fretted and wrung her coarse hands.

Tommy's anxious heart would pound as he watched Mam, her frazzled fair hair severely pulled away from her red and chapped face in a knot behind her head, her intelligent eyes filled with hope that Papa would come out in time to salvage enough of his wages to pay the rent and put food on the table. When Papa didn't come, Tommy's heart would sink. Forbidden to enter the pub herself, Mam would frown, turn blurry-eyed toward Tommy, and send him in to collect what pay he could from his drunken Papa. Tommy hated this chore. Often, he returned empty-handed. Often teary-eyed. Not because of embarrassment of having to beg Papa for money in front of his mates, but for having to see the desperate look in his mother's blue-green eyes when she couldn't feed her children, knowing they would go to bed hungry, their stomachs growling.

"Please, Papa, don't go to the pub tonight!" Tommy's teeth were clenched in frustration as he watched the line of miners begin to thin. His heart pounded in his throat. His hands began to sweat. His stomach churned. Staring stone-faced out the window, he leaned his forehead against the cracked window pane, frosted white from his short, halting breaths. "Please, Papa!"

When Papa was not among the last of the stragglers, Tommy refused to cry. With pursed lips, he turned to his mother, standing at the hearth dressed in her worn-out house dress and apron, stirring the special rabbit stew in the cast-iron pot with increasingly harder jabs of her wooden spoon. She was pretty once, or so they say, but tonight she looked so tired, worn, and haggard.

"Where is he, Mam?" Tommy asked with a quaver in his voice.

Martha returned his anxious gaze. "I'm sure he'll be along shortly," she answered weakly.

Tommy saw the fragile edge to her confidence, her knitted brows, and the tight line of doubt on her lips. "I miss Grand," he murmured with a sniffle, giving in to the fragile child within.

Martha's eyes mirrored his somber expression. "I'm so sorry, Thomas. Papa misses his Papa Grand too, more than he lets on. Sometimes he just loses track of time."

"Is that why Papa never comes home anymore? Because Grand was killed in the mine? Does drinkin' with his mates make him feel better? Better than spendin' time with us?"

"He's hurting in his own way."

Tommy may have looked the spitting image of his Papa, Joseph; tall for his age, thick, wavy hair, dark eyes and handsome, but he was not at all like his amiable, care-free papa who took life as it came, without judgement. Unfortunately, Tommy shared Martha's more weary disposition, which seemed to burden her with all the worrying for the family.

"You know how your Papa is . . ." Martha trailed off. "But his heart is in the right place."

"Yeah, I know how he is."

"He doesn't mean to disappoint you," she said in a concerned voice. "Sometimes it's hard for him to . . ."

"Please, Mam, don't," Tommy interrupted. Tommy didn't intend to be short with his mam, but on this night he couldn't seem to help it.

"Oh, Thomas." She reached over and put a hand on his shoulder, wiping at her eyes with her other hand. "Your Papa loves you."

Sullen, Tommy pulled away from her and turned back to the window. He drew another deep breath, held it in, and stared into the dark shadows of the cold moonlit night. The thought of Grand not being here left him with an ache in his stomach. Reaching into his pocket, he rubbed fiercely at his priceless rabbit's foot for luck, a gift from Grand. His labored breath shuddered out. He suspected all the miners had either passed by or were off to the pub.

"Tommy, all right?" Tommy's two-year-old brother Georgie asked as he crawled up on his lap. Georgie, the youngest in the family, was the mischievous one. If there was trouble in it, he always seemed to find it.

Tommy patted his little brother's arm. "I'm fine, Georgie. Just waiting for Papa."

Reassured, Georgie curled up on Tommy's lap, his thumb lodged securely in his mouth, his free hand reaching up to its usual resting place on Tommy's warm neck. It was a habit Tommy had learned to accept as each night he and his littlest brother slept intertwined with one another on straw-filled sacking on the wooden floor.

As Georgie settled into sleep on his brother's lap, Tommy glanced one last time through the windowpane. The full moon appeared between fast-moving clouds dragging in shadows of rain. In the distance, he saw a lone man with a familiar gait sauntering into view. The man turned his collar up against the rain.

Tommy pulled his tattered sleeve over his palm. "Papa, if you come home to me just this once, I'll never ask you for anything ever again!" He wiped the fog clean from the window glass to get a better look.

Straining to see, Tommy blinked twice. Slowly recognition began to settle into his chest. He exhaled the long breath he'd been holding, and the tension in his clenched muscles began to relax. Relief swept over him. A calming smile eased across his face.

"Papa is coming home!"

⊱ • ⊰

Papa stepped through the front door, his broad, hunched shoulders filling the doorway. With white teeth flashing through cracked lips, his infectious grin spread wide across his lined, coal-blackened face. He bent down and with his powerful arms reached out to his children. His hands were scarred and gnarled, coal dust permanently lodged under his broken fingernails. A badge of honor for a lifetime of hard labor. Tommy's little brother Joey and giggling little sister Edie rushed their Papa in a torrent of laughter. He returned the greeting smiling, with gonna-get-ya hugs and kisses.

Trying not to disturb Georgie, still fast asleep on his lap, Tommy laid an arm across his little brother, edged away from the window, and leaned back against the icebox. His smile brightened as he watched Papa's larger-than-life persona fill the room with laughter and rowdy play. Towheaded Joey, four years younger than Tommy, followed Papa's every move with fascination, his eyes dancing in delight. Joey had the same brown eyes and olive skin, as did Edie and everyone else in the family. But then most all who worked the mine resembled each other after generations of working together, marrying each other, and raising families where every mother seemed to share some in the upbringing of all children in the village.

The village and pitheads of the mine were bordered on the east by Silkstone Common and Nabs Wood. To the north, farmlands and canals for transporting coal stretched four miles to the bustling town of Barnsley. And to the west and south, the vast landholdings of coal king Lord Fitzwilliam, whose Woodhouse Mansion many considered the grandest in all England. Woodhouse was plotted on a seemingly endless estate of manicured lawns, babbling brooks, ponds, and virgin woodlands. Some said the enormous mansion needed only drawbridges and dungeon keeps to evoke visions of castles in earlier feudal days. Stretching westerly for miles all the way to the Peak Mountain Range and southerly as far as the eye could see, Lord Fitzwilliam was landlord and enforcer of laws as only he saw fit. With a velvet covered iron fist, he bent to his will the mine operators, tradesmen, and miners alike.

"Yer all gettin' so big," Papa's jovial voice boomed. "There ain't no more room for ya—squeezed in together shoulder-to-shoulder like pinned-up rabbits." Papa chuckled. "Ya cain't sneeze around here without our next-door neighbor offerin' a hanky."

Without looking up, Martha chided him. "Really, Joseph! You'll have the neighbors pounding on the walls!"

"Yer right there, my dear. These walls are so thin, the business of the family becomes the business of the neighbors," Joseph said, echoing Martha's admonition to his bright-eyed children, who now included Tommy and Georgie, who had joined their brother and sister in play. All were entirely entertained by their Papa. "So, watch your language, me little ones! 'Cause I ain't wantin' ole Mrs. Straycoff a-comin' over here gossipin' to your Mam about it."

As Tommy watched Papa's pantomime of Mrs. Straycoff appearing at the door, he held his stomach, overcome with laughter. The image of their neighbor wagging her finger at Mam caused his younger siblings to dissolve into uncontrollable giggles.

"Next thing ya know, we'll all be in trouble with the landlord," Joseph warned. "They'll be raisin' our rent, callin' in our debt, and your Mam'll tan all our hides. Then there won't be no money for Mam's hard rock candy, will there?" He laughed as he stalked around the room with claws out, to a chorus of boos from his children.

"We be darned lucky if ole Lord Fitz don't require a toll to enter heaven's gate on Judgement Day, so's ole Saint Peter can give us the what-for!"

While the children giggled and rolled around the floor, Martha busied herself at the hearth with final preparations of Tommy's birthday dinner. Tommy looked on at Mam with pride. He had always thought her dexterity at the hearth was a sign of a gifted cook. Balancing the pot, she wielded the wooden spoon with her free hand and was able to pump air into the fire and avoid burning herself on the red-hot iron grate or catching her dress on fire in the flames—by itself a difficult task, but she did it with flair.

Papa leaned over and kissed her on the cheek with a wink to Tommy, who smiled, appreciating the affection shown and received.

Still not looking up from her cooking, Martha rolled her eyes, not quite won over yet. "Joseph, you're all wet, still filthy, and very late. And you smell like a polecat dunked in rum!" But Mam couldn't stay angry long. She playfully brushed him away with a sweep of her arm. "If you would be so kind, please go wash up your hands and face outside."

The children had already had their Saturday night baths, passing through the wash bin oldest to youngest using the same bath water. By the time it was little Georgie's turn, the water had been so dirty Georgie lost the bar of Mam's lye soap in the mucky, charcoal-colored soup. After their baths, with the help of a couple wooden boards, the washtub had been converted back into the dinner table.

"Don't dawdle, Joseph. The children are waiting for the festivities to begin," she warned, brushing her frizzled hair from her face. Then looking up from the sputtering hash with an encouraging smile, she added, "Dinner will be—" she paused with a frown. "Joseph, sweet Jesus, what has happened to your eye? It's all red and swollen." Martha, her brows knitted together, inspected his eye with a squint.

Tommy had not noticed Papa's eye with his blackened face, but now he did.

Joseph's demeanor dulled just a bit. "I was in the wrong place at the wrong time, that's all," he whispered reluctantly. "I got 'tween a bloodied wizened old miner and Boo Black." There was frustration and impotent bitterness in his voice.

The name "Boo Black" sent a chill running down Tommy's spine. It turned his stomach inside out. He knew the rumors of havoc wreaked by Candyman Boo Black. Tommy had seen him once, when Boo had disrespected and threatened his mam. Some boys said he'd even been known to kill miners before.

"We'll talk 'bout it later," Papa whispered to Mam.

Concern filling her eyes, she reached up with her hand to inspect the swelling of his cheek.

"Later, Martha," he reiterated, gently pushing her hand away. Then turning back to rejoin his little ones who were tugging at him from all directions, he bellowed, "Whatta ya tryin' to tell me, Joey?"

"Papa, this ain't no muskrat-guts soup, it's Tommy's favorite," Joey announced in a flourish.

"Yeah," Edie interjected, "tubal greens, hot buttered corn, sweet Ennis hash, and real rabbit stew." Then, with hands together to funnel the words into Papa's ear, she whispered, "Mam bought a whole rabbit!"

"Oh, did she?" He looked at his wife sideways, one eyebrow cocked. "Wit' what money?"

She offered a wan smile. "I've saved a bit of birthday money for tonight. Now, go wash up, Joseph," Martha insisted with raised brows and down-turned lips, rubbing the smoke from her eyes with the back of her hand holding the wooden spoon.

"I'll be back to get ya, me little ones," Papa said as he stepped outside into the cold rain. He headed down the alleyway to the neighborhood pump to wash off as much of the grime as he could.

Steam rose off the simmering hash, and the soup bubbled wildly on Mam's coal-fired stove. Intoxicating smells filled the room. As rain poured down, the clapboard walls began to shudder, the door rattled, and the rusted tin roof clattered, all while little Georgie, Joey, and Edie noisily rolled about the shanty floor.

Even with all the commotion, Tommy didn't hear a thing. He couldn't stop imagining what he'd do to Candyman Boo Black someday for hurting Papa and disrespecting Mam. *One day,* Tommy thought, *when I grow up, I'm gonna make him pay—no matter how big and scary he—*

". . . Thomas! Thomas, d'ya hear me, son?" Joseph boomed, interrupting Tommy's thoughts. "Why d'ya not listen to me, boy?" Staring down at *Robinson Crusoe* on Tommy's lap, he didn't wait for a response. "Ya should be helpin' your Mam."

Tommy looked up, oblivious to the table being set for dinner. "Sorry, Papa," Tommy answered in bewilderment. "I'm sorry, Mam."

Joseph took the book that still lay open in Tommy's hands. "Yer always lost in another world 'tween the covers of some book. Not payin' enough attention to the world yer in." Joseph thumbed through the pages.

"But Papa?"

"No *but*s."

"What can I do to help, Mam?" Tommy asked, trying to escape the irritation in Papa's voice.

"You can set up for dinner."

Edie piped up. "Papa, you wouldn't believe the big books Tommy reads now! You should see 'em. I'll bet he's the best reader in the whole village. Maybe even in the whole world!"

"Hmm."

Joey joined in. "Mam says he could go to London someday and become an engineer or a scientist."

Joseph looked at Martha with exasperation. "She does, huh?"

Martha didn't look up. She seemed to glower harder into the sputtering hash in the hearth, practically beating it to a pulp with her wooden spoon.

Tommy frowned, his jaw tightened. "No, I'm not!" he blurted out with conviction. "I'm gonna be a coal miner, just like Grand and Papa. Startin' Monday mornin', I'm joinin' the rest of the boys down in the mine."

Martha turned to her son with hurt in her eyes. Then, frowning, she met Joseph's eyes with a cold, steely stare. Finally, she looked away, closed her eyes, and steadied herself against the mantel.

Tommy's heart sank. He was almost sorry he had blurted out his intentions.

"Martha?" A stern-faced Joseph glared at his wife. "The Wrights been workin' South Yorkshire coal mines for more 'n a hundred years. I started when I was seven, Grand at five. Me boys'll be doin' the same," Joseph said with a great sense of hurt pride in his voice.

A sober Martha looked down at her little boy, her eyes rimmed with scarlet, her cheeks puffy and pink.

"It'll be okay, Mam!" Tommy said. Everyone in the village had accepted his future—everyone but Mam. Tommy felt sad for her.

As much as Papa was malleable under the company rule, Mam was fiercely determined not to be molded into expectations that would define her role in the family and in village life. Martha, who had once been a housemaid to the countess at Woodhouse, often chafed at the restraining control, especially when it came to her children and their education. She was born under a different star, with a wide streak of intolerant defiance in her backbone. Tommy remembered what Grand had told him about Mam. "She'll stand up against the avalanche of life for what she thinks is right. Like all women in this here village, she'll be beat down for her beliefs. But I suspect she'll go down fightin' to her last breath. Only your Papa can calm the fire in her gut."

"Things are what they are, Martha," Joseph whispered, trying to console her. "And we ain't never gonna change it."

"Why not?" Mam turned to face Papa, hands opened out pleading to him. "Isn't it a good thing to encourage him to hope for a better life?"

"I love ya, hon. But what is, just is. Ain't nothin' a body can do about it," Papa said with compassion in his voice. "There ain't no use in complainin'. Our boys'll be workin' for the company as coal miners the whole of their lives."

"But what about opening up opportunities for our children's future?"

"We only get one chance at life, Martha. We can't change the way things is."

"One chance may be enough if we give them what they need to seize the moment—if and when that moment comes." Mam responded with such passion and hurt in her voice it made Tommy feel awful. He stood in wonder at his Mam's ferocity.

"Hon?" Papa took in an exasperated breath and blew it out in frustration.

"If times are ever to change for our children, we have to help them," she continued. "Shouldn't we give them the education and the courage to make that chance possible. Shouldn't that be our goal?"

Frustrated, Joseph tried again. "We don't want to be fillin' their heads with false hopes now, do we, love? We have to make the best of things the way they are. The way they've always been."

"I'm sorry, Joseph, but I want more for them than that."

Tommy licked his dry lips, wiped the back of his hand across his dampening forehead, and tried to push back on the anxious knot inside him. His Mam's words touched a chord in him. He shared her passion for learning, for reading of faraway places, for stories of science that would change the world outside of their village someday. He had tried to put it out of his mind. Still, he couldn't help but wonder what it was like in London, Paris, America.

"It's not the proper order of things," Joseph insisted in a whisper.

"But Joseph—"

"No buts, Martha. Please!" Joseph pleaded. Then he turned and pointed his finger at his children. "Who can tell me why we're celebratin' here tonight?"

"It's Tommy's seventh birthday," came the chorus. Tommy pushed out his chest with pride.

"That's right. It's a big day for our Thomas." Joseph looked on with pride at his eldest son who, dressed in his only high-collared shirt, reflected a boyish blush of proud embarrassment. "It's the day he starts to becomin' a man and contributin' to the family income," Joseph confirmed. "We're all proud of ya, son."

"Thank you, Papa."

"Okay, Joseph," Mam conceded in quiet acknowledgement.

Tommy would leave the nurturing bond of her schooling, his books, his life of learning under her protective wing, to begin his new life working down in the pit, just like all boys in the family had done for more than a century. A lifetime of blasting jagged arteries and digging airless capillaries through veins of coal buried deep in the bowels of the earth in support of England's industrial revolution. Martha's son would mine a vein of black gold that brought untold wealth to its operator, Sir Robert Clark, and landlords up and down the seam like Lord Fitzwilliam, who owned not only mines and villages but the stores, the iron works, the shanties where the workers lived, even the food they ate and the mortgages on their lives.

"But," Martha whispered as she finished preparing dinner, "I won't let Thomas's education go to waste. With God as my witness, I promise you this, Joseph Wright: I have not given up. I will do everything in my power to ensure Thomas and the rest of my children leave a worthy footprint of an honorable life well lived. An educated mind cannot be enslaved by men like Lord Fitzwilliam."

☙ • ❧

When dinner was over, Martha washed and dried the dishes while Joseph sat in front of the fire, telling stories about the week's events in the mine. Tommy sat right in front, laughing and rolling on the floor as he listened to his Papa's animated interpretation of the actions of his mates. It was the best of times when Papa was home.

"It's time for scriptures and prayers," Martha interrupted when she finished her work, much to the dispirited dismay of her four children.

"Really, Mam? Papa is home. We don't want the fun to end," Tommy argued on behalf of all the children.

"Well, I think that is a good reason to prevail upon your Papa to come home more often."

Joseph, a bit irritated by the undercurrent in her words, nevertheless respectfully bowed to her wishes.

"Well, that's enough for tonight, me little ones," he declared as Martha wiped her hands on her apron and sat down in her usual place in front of the fire. Her little brood obediently gathered around her, all bathed in firelight.

The ceiling, corners of the room, and everything else outside their family circle seemed to recede into flickering shadows. Tommy frequently noticed that when Mam opened up her Bible, the brightest light always seemed to shine on her. That night, she read from the book of Matthew while Tommy and the rest of the family listened to her melodic voice. When she finished, she looked at each child, and they all clasped their hands together without further direction.

"We'll be havin' our prayers now, will we?" Joseph asked, not used to spending his evenings with the family.

"Yes. I think it's time for prayers, then bed?" Martha smiled up at her husband and Joseph reluctantly took his place beside her.

Tommy leaned his forehead onto his clasped hands, which were strategically placed atop his folded knees so his mam could not see his face. If he drifted off into daydreams in this position, he reasoned, it just might escape her notice, though she seemed to have an uncanny sixth sense about these things.

The words she spoke went unattended by his ears, but her voice, lulling and soothing like a gentle breeze, as always touched his heart. As far as Tommy was concerned, if his mam spoke it to God, it was gospel. In Tommy's mind, his mam's actions were close to godliness in all things—except, of course, on the subject of coal mining. When she thanked God for the health and strength of her family, a deep sense of quiet, peaceful refrain fell upon him. A calm filled his heart. He knew, above all else in life, Mam loved him, and he felt certain that if anyone could commune with God on his behalf for his many sins, it was her.

When Tommy listened to his Mam pray, he imagined Grand looking down on him from heaven. He wanted to make Grand proud of him. It was time to go down into the mine with Papa. Time to do his part for the family. He closed his eyes, wondering just what it was going to be like to put his schooling and all else aside, to take up the family business.

Chapter 2

꜕ · ꜖

TOMMY SAT ON THE BENCH at the washtub table, arms folded over his knapsack, feet dangling above the floor, waiting on the edge of the day for his papa to tell him it was time to go. He could hardly believe he was finally going down into the mine with Papa and the other boys in the quarter. He had risen extra early that morning so he could pump the sulphur-tainted water from the well at the end of the block, finish the rest of his chores, and even go to the loo before the queue started along with the worst of the stench. He had prepared his knapsack and sat ready at the table—all before Mam had finished preparing breakfast, which he'd already finished eating.

"We're so proud of you, Thomas." Martha absently reached over the table to fix her eldest son's collar. "You're a big boy now." She gave him her warmest smile.

"Yes, Mam." Tommy tried his best to pay Mam the attention due. He didn't want to be disrespectful, but this was his morning to be with Papa, and he could hardly contain his excitement. His eyes darted around the room, keeping close track of Papa's every move as he dressed and prepared his knapsack for the day's work. "It seems I've waited me whole life to be a coal miner."

"Proper English, Thomas," reminded Martha. Ever the tenacious taskmaster, she had been hell bent on using the few years she had her son entirely to herself to prepare him to withstand the forces of change outside their family home and the challenges he would have to face. But his formal education was over now.

"Yes, Mam."

Tommy fidgeted. His mind was racing. His feet tapped on the floor and his fingers drummed the table. He would meet other boys today, make his own money, and discover his own stories to tell by the fire.

Martha considered his impatience. "You'll have responsibilities every day in the mine," she reminded him. "But besides learning the skills of a coal miner, you must learn to study things out for yourself, stand up for the right, and continue with your studies when you can."

"Yes, Mam."

"I know you'll do your best, Thomas. You're such a gifted boy. It's a great blessing," she said with conviction, "but remember, where much is given, much is expected."

"I won't forget, Mam." Tommy looked up at her.

Until now she had always been his unyielding guidepost. Knowing she would always be there for him brought comfort. He felt a warmth wash over him whenever he heard the soft sound of her voice, felt the kindness in her touch, breathed in her sweet, familiar smell. Somehow, Mam's quiet solemnity always made him feel safe, secure, and loved. He would miss her and knew she would miss him. It left a hollow feeling in his stomach knowing that he must learn to be apart from her.

"You'll be adding to the family income." She looked deep into his eyes. "But . . ." She turned away to wipe at her eyes.

"I promise, Mam. Just like ya taught me," Tommy confirmed, feeling guilty for wanting to break away, to get on with this new adventure.

"Just like *you* taught me," Martha corrected reflexively, reaching over to tuck in Thomas's errant shirttail. "And there is still a lot more to learn isn't there, Thomas?"

"Yes, Mam," he nodded. "Can I go now?"

"*May* I go," she corrected. "And not quite yet."

Martha reached behind her, picking up a small, beautifully bound book and a leather notebook.

"I bought you these with a bit of money I've saved for this occasion. It's to celebrate your first step in growing into a man."

"Oh!" Captivated by the two books Mam held in her hands, he forgot about Papa for a moment. He reached for the first, a novel. He thumbed through the pages, hefting its weight and passing a hand admiringly over its cover.

"It's called *Gulliver's Travels*. It's an adventure story written by an English author."

He reverently skimmed through the pages of the book.

"And you'll love it. It's for you to read by candlelight during those quiet times in the mine. It will make the long hours go by faster."

"Thank you, Mam," his eyes brightened. "Wow, it's so pretty. I can't wait to read it!"

Martha handed him the leather-bound notebook. "This second book is a journal for you to write in," she said as he flipped through its pages. "As you can see, I've printed on the outside cover *Thomas Wright, South Yorkshire England, 1837*. The pages are blank. It's up to you to fill them with stories of your own experiences. We can review them together from time to time, if you'd like. Or maybe share them with your brothers and sister."

"Thank you, Mam."

"Here, let's put these in your knapsack. It'll be our little secret, all right?" Then she whispered, "Now, you go and do your best.

Remember, Mam and Papa are always here to help you." Gently, she reached over and touched his cheek with the warm, calloused hand of a hardworking mother. "I know you're a big boy now, and brave too, but sometimes you'll be afraid. You must believe in yourself. Only you can make the choice to overcome your fears." She paused to allow the message to sink in. "And always remember, the courage to do the right thing is the most important of all your choices, isn't it?"

He nodded his head. "I'll remember . . ." he trailed off when he saw she was upset. "Are you okay, Mam?"

"Don't you worry." Martha smiled through the moisture in her eyes and ruffled his hair. "It's just what mams do when they send their favorite little men off to work in the mine for the first time!"

Tommy reached out and held her hand. She looked away and, with her other hand, brushed down her apron. He didn't like seeing Mam this way, his unbendable pillar of strength.

"Ready to go?" Papa looked to Tommy as he moved toward the door. The sun would not rise for another hour.

Bleary-eyed, both Joey and Edie had dragged themselves out of their warm beds to give Tommy a hug goodbye.

"Ya gotta promise to tell me 'bout it tonight," Joey said, still rubbing sleep from his eyes.

"Me, too!" Edie chimed in. "Bring me a present, okay?"

Tommy rolled his eyes. "I'll see what I can do." He still didn't quite understand the mysteries of what went on in little girls' minds. Then he leaned down to kiss little Georgie goodbye.

"Tommy, don't go," Georgie pleaded as he wrapped his arms tightly around his big brother's neck.

"Don't worry, Georgie, I'll be back." Tommy roughed up Georgie's hair, then slid out of the grasp of his little brother. With droopy eyes, Georgie stuck out his lower lip in a frown and stared at Tommy, who, still down on one knee, looked into Georgie's eyes.

"Can't I go with you?" Georgie begged, used to shadowing Tommy everywhere he went.

"Sorry, Georgie. Not this time. Today it's just me and Papa."

Georgie scowled at his father. "I don't like you, Papa."

"But tonight, we can play in front of the fire when I come home." Tommy patted his little brother on his back. "And with the first money I make, I'll buy you a bit of sweets. How's that?"

"Okay, Tommy," Georgie conceded with downcast eyes. "But I want curlicue coal candy."

"Okay." Tommy giggled. "I'll miss you." Then Tommy stood. "I'll miss you all." Then he turned and gave Mam a big hug. "I love you, Mam," he whispered into her ear.

"You be safe."

Papa beckoned from the doorway. "It's time, Thomas."

With a reluctant smile, Martha nodded toward her son. "What must we always remember?"

"Be good, do good!" And with that Tommy turned and followed his papa out the door.

≯ • ≺

"Hey, Tommy! Ya ready fer some collierin'?" Zac, Tommy's older cousin, came out of the darkness to join Tommy as he walked toward the mine. Papa, well out in front and engaged in conversation with another miner, took no notice of his nephew. At eleven, Zac had a strong back and shoulders from the heavy work of a Jenny-boy hooking up trains of carts and lowering them down the coalface.

"Watch out for dem mud snakes, Tommy!"

"Mud snakes? I never heard of no mud snakes," Tommy responded, falling farther behind Papa as he slowed his pace and fell in beside cousin Zac and his younger brother Abe.

"Oh, yeah! They be deadly," Zac added with a tight-lipped expression, his brows furrowed in all seriousness. "They put all the new boys in the dark where the snakes live."

Tommy's face went pale. "Well, whatta ya do?" His imagination wrenched into perilous thought, he wiped the back of his hand across his dry lips and looked at his cousin with wide eyes. "Why do they put the new boys—"

"Well, I suppose," Zac went on, "'cause they figure iffen we gonna lose a boy, it might as well be a young'un who ain't worth much to the company."

"Oh." Tommy's heart began to beat rapidly at the thought, a concerned ache tightening in his chest. He didn't want to consider the consequences of being bit by a mud snake.

"Not much a body can do, and—"

"Don't listen to 'em, Tommy," Abe interrupted, slapping Tommy on the back. "He's makin' it all up. There ain't no mud snakes."

"Really?" Relief rushed over Tommy as he let out the long, rasping breath he'd been holding and joined his cousins' laughter at the teasing.

Zac mussed Tommy's hair. "Nice to has ya with us, Tommy. Any son of Uncle Joseph, da best miner on the seam, is surely welcome as a colly."

On this cold morning, in the shadows of the fading night sky, Tommy strutted and skipped good-naturedly with his cousins toward the mine. It was a time of innocence, and despite their longing to grow up, most were still not yet ready to become men.

Joseph dropped back from the other miners. "Zac, Abe, how's yer mam doin'?"

"Fine, thank ya, Uncle Joseph," Zac offered. Both Zac and Abe tipped their hats and moved on down the pathway. Their papa had been killed in the same explosion that killed Grand last spring. They were now the sole bread winners for their mam, two little sisters, and the new babe.

Joseph pulled in beside Tommy in the strung-out line of miners. "Ya knows what you'll be doin', son?" Papa asked as they walked briskly through the gate into the Silkstone yard.

"Not exactly, Papa," he answered, trying to keep pace with his papa's long legs.

"Ya'll be makin' yer livin' as a trapper."

Tommy's forehead creased in curious wonder. "What do they do?"

"They ventilate. To keep the air movin' sos the other miners can breathe down there. Sos the gases don't build up and start a fire—or worse, an explosion."

Joseph paused to let the message sink in.

"See those buildings over there?" Papa nodded in the direction of two large brick buildings near the pithead entrance to the mine, still a short distance in front of them.

Tommy looked at the one building adjacent to a stairway leading up to a wooden platform surrounding the mine shaft. Mounted atop the platform was a giant, twenty-foot-diameter spool of cable, which led up and around a pulley stationed on a timber structure high above the platform. Hanging from the end of the long cable was an iron cage full of miners ready to be lowered down the shaft into the mine.

"The big brick building yer lookin' at has a steam engine inside that runs the winding gear fer the cable that lowers the cage and men down into the mine," Papa explained. "The building next to it pumps air deep into the tunnel drifts pushin' out the gases. When the trapper opens the doors to the drifts properly, all goes well and fresh air pulls off the gases. Iffen he don't do his job proper like? Big problems!"

Tommy wasn't exactly sure what Papa meant by "big problems," but he didn't want to reveal his concern by asking too many questions either. His hands began to sweat, and his heart began to skip a beat. He was becoming concerned that maybe there was a bit more to being a trapper than hanging with the other boys.

Without another word, Tommy fell in line behind Papa, staring at the ground, trying to keep up, as they walked the remaining distance through Silkstone's equipment yard. Furrowing his brow, he concentrated hard on his task ahead.

"What if I make a mistake, Papa?" Tommy asked as he started up the steps to the platform.

Papa turned back. "Let's not be thinkin' like that."

"Yes, Papa."

This daunting revelation was upsetting. To avoid further revealing his anxiety, Tommy squinted down at his feet and hurried up the steps to the loading platform. He rubbed his sweaty palms on his trousers and silently prayed for a little help to not make a mistake.

In nervous distraction, Tommy walked across the loading platform without noticing the rushing sounds of the winding gears of the steam engine, the clanking of the spool of cable, or the scent of mechanical grease in the air. He didn't hear the whining of the cable traveling up and over the pulley above, nor did he notice the carbonic stench and taste of coal dust rising up with the breeze from the mine shaft. He put his sweaty hands in his pockets and absently followed Papa past the bankman at the disembarking platform and on to the iron cage that would take them three hundred feet down into the pit.

In the shadow of early morning darkness, Papa stepped from the loading platform through the open door and onto the floor of the iron cage. A loud *screeeeeech* of steel-on-steel jolted Tommy back into the moment. Rattled, he reached out for the cage dangling at the end of the long wobbling cable, but it swung away from his grasp and slammed against the opposite wall of the mine shaft with a loud *whack*. Tommy jumped. His pounding heart rose into his throat. His shallow breathing stopped. The clanging of the oscillating iron cage, squealing of cables, and grinding of the winding-gear engine froze him in fright.

He looked down. The light from the swinging cage lantern had swept away the darkness of the widening gap between the cage and the edge of the disembarking platform. The illumination from the flickering light telescoped deep into the darkness of the cavernous shaft. His confidence shaken, Tommy stared into the disappearing abyss. His forehead broke out in cold sweat. His queasy stomach churned, acid rose into his throat, and he retched up his breakfast.

Woozy and faint, Tommy's knees went weak. He made another hapless swing of his arm at the cage, but it was too far away. His knees buckled. He lost his balance, and he fell forward toward the ever-widening chasm of darkness.

Suddenly, like a lightning strike in a summer storm, a hand shot out from the cage and grabbed Tommy's outstretched wrist. With one powerful, determined pull, Joseph yanked his son aboard the swaying cage full of miners. Tommy threw his arms around Papa's waist, burying his face deep into his stomach. Shaking in fear, he tried to bring his emotions under control, to retain his dignity, but still the tears came.

"It's all right, Thomas! Don't look down now. Always keep your eyes fixed straight ahead when yer steppin' into the cage."

Joseph pulled the cage door shut, knocked two times on the metal frame with his hammer, and the cage began its descent. Tommy's muffled sobs, buried in the security of Papa's strong arms, were drowned out by the clanging of the cabled cage as it lowered into the pit.

"I'm gonna be strong for my papa. I'm gonna be strong. I hope I'm gonna be strong. Oh God, please help me to be strong!" he whispered over and again as down, down, and farther down they went into the darkness.

The panic began to ease, but Tommy held on tight to keep his hands from shaking. The only audible noises were the creaking cable rolling over the pulleys and the *clang, clang, clang* of the cage against the rough-hewn rock walls of the mine shaft. As they lowered deeper into the bowels of what seemed the center of the earth, the stuffy air warmed, smelling of coal. Water seeped in from all sides of the shaft, soaking the cage and dripping down on him. After what seemed an eternity, light began to reappear in flickering shadows. The cage jerked to a stop. At Papa's encouragement, Tommy released his white-knuckle grip from his belt. Joseph opened the door of the cage, and with uneasiness, Tommy stepped out onto the disembarking platform and stood, waiting for Papa holding the door open for the other miners.

As each of the other men stepped out, one by one they respectfully nodded to Papa, "Thank you, sir!" doffing their caps, some giving Tommy an encouraging pat on the head.

Papa shut the door behind them, banged three times on the steel frame with his hammer, and the cage lifted back up the shaft.

"Yer Pap is the best miner in these parts," offered one miner as he harnessed up, securing his pick, shovel, and stout pole peggie. "'Tis in his blood."

"Pay close attention, and your Papa'll teach ya well," said another, grabbing his oil lamp in final preparation for the day's work. "He's the best fire boss in the mine."

It was clear that Papa was in his element here, three hundred feet below ground, where he had lived most of his life. He exuded a quiet confidence, reciprocated in earned respect and friendly banter from his fellow miners. There would be no ribbing for Tommy's tears on his first day of the rest of his life in the mine.

Chapter 3

>· ≺

TOMMY HURRIED TO KEEP PACE with Papa. They were in one of several widening tunnels spinning off in different directions from the disembarking platform like spokes on a wheel. Each led to smaller chutes, or drifts, where each miner or pair of miners, sometimes husbands and wives, continued on their way until reaching their respective monkey-head diggings. Being three hundred feet below ground was far different than Tommy had imagined. All sound faded into silence as they headed farther away from the cage. The temperature was warm, and his shirt soon clung to his damp skin. The dank smell of decaying mule dung, human excretions, and coal dust, musty and putrefying, hung so thick in the air, Tommy thought he could taste it.

Tommy took in as much as he could in the shadowy darkness. For the moment, his fears were overtaken by curiosity. The flickering shadows of sparsely placed candles along the widening tunnel played tricks on his senses. His hurrying feet tripped and stumbled on the shadows of the uneven stone, causing him to fall behind Papa. Anxiously, he leapt into a trot to try to keep up.

"I'm carryin' on the proud tradition of me family," he muttered to himself, echoing Papa's admonition. "I promise, Grand, I'll be strong." He took deep gulps of air to fend off his anxiety. He knew it would devastate Mam if he let her know how scared he was—as much as he wished he could. In that moment, he promised himself he would never tell her of his fears.

"Come on, Thomas, thisaway! Our monkey head be over thisaway."

"I'm coming, Papa!" he called out.

Finally, Papa stopped at a trapper door. "This here's Katie Garnett," Papa said, smiling at Katie, who was maybe nine. In her shyness she said nothing as she pulled open the trapper door and a surge of warm, acrid air came down the drift passageway. "She be a trapper too."

Tommy nodded and offered, "Hello." And she returned the courtesy.

Papa took off his jacket and stuffed it into his knapsack.

As soon as Tommy had done the same, Papa dropped to his hands and knees and began crawling into the thirty-inch-square opening in the wall.

"Follow me, Thomas," he called back. "This be our drift to me monkey head."

Tommy hesitated for only a moment before dropping down to his hands and knees to follow, but by then Papa was already well down the drift, disappearing into the darkness.

"Papa?" a reticent Tommy called out, hurrying after him. The air in the confined space was stifling. He bumped and scratched his head on the ceiling.

"Watch your head, son. Keep your hands on the cart rails."

Now Tommy understood why the previous, larger tunnel was called "the widenin' tunnel." As he rushed down the rails on his hands and knees, his knuckles scraped the rough rock floor and his knees hit the crossties. He stopped in the ever-tightening space to check the bleeding trickling down his face, but it was too dark to see.

Joseph's bobbing oil lamp dimmed in the distance. The darkness closed in. Panic began to grip Tommy's heart as it pumped ever faster. He pulled in several labored breaths to regain his composure, then crawled faster, as fast as his little hands and knees would carry him.

"Oh, Papa! Please don't leave me behind," he whispered, teary-eyed, wincing each time his knuckles scraped on the hard rock floor. Involuntary breaths came faster now, and Tommy began to hyper-ventilate. He could not push back the surging wave of acid reflux and rising panic in his throat.

"Papa?" he choked out as he arched his back to keep his eyes on the distant bobbing lamp. Desperate to catch up to the waning light, he ignored the pain, willing himself to stop the tears from coming. "You be a man now," he whispered. "You be a man." He repeated this mantra over and again, grinding his teeth in sync with his shuffling hands and knees. His heart pounded in his chest; puffs of short, erratic breaths came out hard. His wrists and forearms ached, but he dared not stop again.

"Papa, are you there?" Tommy's voice quavered.

"We're almost there, son," came back Papa's answer. "It's best to keep ahold of the two tracks. Just another hundred feet. Are ya all right, son?"

"Yes, Papa!" Tommy's voice was weak and fractious.

Finally, light began to fill the end of the drift. Tommy's vision began to return. His little arms and legs rushed the remaining yards to his papa, where he was surprised to find him in a wide-open room where he could stand. Papa had set down his oil lamp and was lighting a candle.

As panic began to subside, Tommy sucked in a deep breath of the humid air to calm himself. He stood up and looked around. "This is your monkey head?" he asked in wonder.

"Yup. It's where I work," Joseph confirmed, pointing to the seams of coal. "Where I make a living for our family."

The room's ceiling was high enough for a near-grown boy to stretch to full height. Tommy leaned into his aching back, trying to stretch it out and not think of his scrapes and bruises, taking in the many details around him. He wiped the sweat from his brow. Following Papa's example, he pulled out his water pouch and took a drink, then tucked it back into his knapsack.

"It's hot in here." He breathed in deep as he moved closer to his Papa.

"That it is, son."

Tommy looked up to see that narrow cracks and crevices in the stone walls were filled with coal. Each seam split off in different directions.

Tommy peered into the empty tubs in the center of the room.

"The hurriers and thrusters brung 'em in last night," Papa pointed out. "They be waitin' to be filled and pushed back out me monkey head, loaded into the coal carts, lifted outta here and on to London, Manchester, Sheffield, and cities beyond. We do important work here!" Joseph explained with pride. "Coal for folks all over Europe!"

Tommy turned back to see Papa stripping down to his skivvies, blackened as dark as the coal he'd been digging. His eyes widened. "Papa . . . what are ya doin'?"

Joseph chuckled, putting his clothes onto a spike in the wall by his knapsack. "When ya be workin' down here, the hot air and gases off the coal seams makes it quite toasty."

Tommy stood speechless, his eyes wide, his mouth hanging open.

Joseph chuckled. "Close your mouth, son. Ya'll be catchin' flies. These here be me workin' clothes. Ya can see why I can't wear 'em home," he snickered. "Maybe I should say ya can smell why. Yer Mam would have me hide."

"Ya don't even wear your shoes and clothes? Yer naked but fer . . . ?"

"No need, son. Not down here when yer workin'."

Never imagining this was the way his papa worked every day, Tommy couldn't take his eyes off him. There were knots of calcium deposits on his knees and the tops of his feet, and calluses from years of working on his hands and knees.

Joseph struck a match and relit his oil lamp. "I'll be a-diggin' this here slot with this pick and shovel. Come, let me show ya how it's done."

Papa slid into a slot cut in the wall about the size of a coffin. Using a steel pick on a seam of coal above his head, he began hammering. Papa's eyes were closed; he worked by feel with an experienced hand. Chunks of coal splintered off in every direction, bouncing off his arms, chest, and legs. Puffs of black coal dust filled the space as he hammered away.

This dirty work continued until Papa, soaked in sweat, had dug out a horizontal slot along the base of the floor. Now it was twice the size that he had started with. Joseph stepped out, wiped the back of his hand across his face, smearing the sweat and coal dust, and stood beside Tommy admiring his work.

"That's how it's done. A bit messy," he laughed. "But when I undercut the face and sets the charge, it'll bring down the whole thing without wastin' much blastin' powder."

Tommy stood quietly, trying to take it all in. An eerie chill ran up his spine while he envisioned Papa sliding into the slot in the wall with a charge of black powder and a lighted oil lamp.

"Is the blastin' dangerous, Papa?"

"Gotta be careful."

Is this how Grand was killed? he wondered. Had he dug his own grave and died in darkness under three hundred feet of earth, rock, and coal?

Tommy's stomach churned. Wiping the sweat from his brow with the back of his sleeve, he took another deep breath, pushing back on the anxiety.

"It's a difficult job," Joseph said, "but an important one. Ya'll be learning the tricks of the trade soon enough." Papa watched Tommy a moment, and then asked, "Whatta ya say we gets ya to the trapper post where you'll be workin'?"

☞ · ☜

Tommy was determined to keep up as they headed toward his own workstation, staying in step behind his papa in the shadow of the dim light of his oil lamp. The tunnel narrowed and the ceiling lowered. Water sloshed around his ankles in the dip of the low point in the tunnel, mud oozing between his toes.

"Keep your head down, son, or ya'll be smackin' it up top," Papa said as he bent low, then down to his hands and knees in the mud. Tommy's hands almost drug in the water. The ceiling continued to lower until Tommy had to drop to his hands and knees in the muddy water.

"The drift may seem small and confinin'," Papa said, "but followin' some seams it can get even smaller, sometimes maybe only twenty-two inches high. Ya have ta lay on yer back, nose almost touchin' the coalface. And sometimes when ya pull out the support posts after digging out the coal the drift collapses behind ya—tricky."

Tommy tried to imagine it.

"What's that awful smell?" the boy asked. The sudden abhorrent stench made it even more difficult to breathe in the already confined space.

Papa ignored Tommy's question.

Once out of the water, the slope on the pair of tracks flattened, and the ceiling of the drift opened up higher as they reached a batten timber doorway, supported by two wooden trusses on either side.

"This here be your trapper station, Thomas. When the door closes, it seals off the tunnel for ventilation purposes and to rid the gases comin' off the coalface. See this here fist-size hole?" Joseph brought his oil lamp close to illuminate a knotted rope that stuck through the hole in the door. "This be your trapper hole. Ya stick yer hand in, grab the rope, and pull with all yer might 'til the door swings open and the corve carts pass through. Simple as that."

Joseph took one of Tommy's candles out of his knapsack, lit it, and dripped some wax on a nearby ledge already caked with hardened candle wax. He placed the candle in the soft wax, and the light flickered and danced off the jagged walls, making the tunnel feel eerily haunted. Tommy could now see a stool lodged in rotting mounds of deteriorating garbage and dried human excrement—the source of the awful stench.

"It's a disgusting mess, ain't it?" Joseph said. "Your Mam'd have a conniption iffen she saw it. Now look here, this is what you'll need to be doin'. Listen and feel for the vibration of the rails first, then the sound and sight of the cart followin' behind. When ya hears it, begin openin' the door." Joseph demonstrated, reaching through the hole in the door to grab the knotted rope and pull the door open. "Ya don't wanna wait too long. The carts full of coal could crash into the door when it's a-comin' downhill. That'd be dangerous— coming uphill, the hurriers and thrusters be a-pushin' and pullin' empty carts. Pay attention."

But Tommy was distracted by the small footprints that ran between the trapper doorway and the stool. And then something moved in the messy pile.

"Papa, did you see that?" Tommy pointed as a dark streak flashed between the legs of the stool, scurrying away into the darkness.

"Those be rats, son. Young'uns I 'spec here to greet ya." Joseph chuckled. "Papa rat will be along shortly. They be some big ones." He laughed. "Some more than two-foot tall standin' on their hind legs."

"I don't want to meet 'em, Papa." The thought of being in the dark alone with rats made him shiver. "What am I supposed to do about them?"

Joseph laughed. "Just keep your lunch up on the shelf sos they can't get at it."

"Oh." A skeptical Tommy looked around the stool.

"Rats rule the world down here," Joseph explained. "They been here long afore us and will be here long after us, too. And they'll become as good of friends as any ya'll ever have. They can sense a cave-in, explosion, or thin air long afore we can. So, keep an eye on 'em. When they start runnin', you run right along with 'em! 'Cause it'll be time to go."

"Really?" Tommy didn't want to think about it. "I don't think I want 'em as my best friends." He felt overwhelmed.

Joseph took one of the candles out of his own knapsack and set it on the ledge.

"I'll leave this here one for ya today, son. There'll be none but this one fer yer twelve-hour shift. Ya'll want to blow your candle out from time to time. Save it for when ya really needs it. Understand?"

Tommy nodded. "I'm not afraid of the dark, Papa."

"I know ya ain't, son, but after a while ya'll be wantin' the light, iffen for nothin' else but to use the loo or eat your lunch. I can't spare no matches, so ya'll have to light your candle off of the oil lamp of the hurrier comin' through your door." Joseph pointed to the railway tracks behind the door. "The corve carts will be a-comin' thisaway from the widenin' tunnel when they're empty and t'other when they're full. They'll be layin' track and bringin' coal out from the new diggin's. We be chasin' the coal wherever it leads us. That's what we do down here."

Joseph turned to face his son. "Now here are a couple important things I want to tell ya, so listen real close. If your candle starts to burn real bright, in a flickerin' sort of way, light outta here real fast. Come to me or one of the other miners. Do ya understand, son? This be real important now, all right?"

Tommy nodded, with a frown of concern. "I think so, Papa." He didn't want to say too much for fear his emotions might get the better of him. What would he do if his candle burned out? Was he done for? A shudder ran through him.

Joseph gave a long look at his son, then nodded at the door. "Afore I go, let's see ya work that door, Thomas."

Tommy reached into the hole and pulled on the rope.

It didn't budge.

"Really lean into it now, son."

"Okay, Papa." He grabbed the rope with both hands, put one foot up against the wall for leverage, and pulled with all his might. Slowly, the rusted iron hinges on the bulky door squeaked and squealed until finally, the door swung open.

"Now remember, pay attention." Joseph nodded encouragement. "And no dozing off, ya hear?" He patted his son on the back.

Tommy took a deep breath and held it to help pull himself together, then let it out slowly.

"And iffen ya got candlelight, ya can read your book."

Tommy blinked, surprised.

"Yeah, I know about the book your Mam gave ya. Don't tell her, though. There's only two ways to argue with your Mam, and neither work."

"Okay, Papa."

"A good woman, that one! She be makin' ya far better than your Papa, I suspect."

Tommy nodded, afraid to be left alone. But he didn't want to tell Papa. What would he think?

Papa patted him on the shoulder. "Try not to think too hard about this new responsibility, Thomas." He turned and gestured to the tracks. "On the way back, the cars are empty, so the hurriers and thrusters will want to stop and talk. And 'cause the drifts are low and narrow in this part of the mine, most all workin' here is young boys and girls not much older than you. It's a good break for ya both. Ya can light your candle and make good friends that way. Just don't dally too long. They'll be few older miners in this part of the mine, so you'll be on yer own to do the right thing. Do yer job and all will be well—understand? We not be wantin' any accidents here."

"Yes, Papa."

Joseph put his hands on his son's shoulders. He looked him in the eye. "Thomas, your Mam has taught ya real good with your speakin', numbers, readin', and all. I promised her I'd support ya in that. But yesterday, ya was a slip of a boy. Today yer a trapper, on your way to becomin' a man. It's an important job. Lives depend on you, so do good by it, and in a short while, ya'll move up the ladder—a thruster, a hurrier, and eventually, a miner like your papa."

"Yes, Papa. I'll do you proud."

Papa grinned. "I know ya will, son."

With a wink, Papa turned and headed down the tunnel, leaving seven-year-old Tommy standing bent over at his post to avoid hitting his head on the ceiling. Tight-lipped, his knees trembling, Tommy took in a shaky breath and watched the light from Papa's oil lamp rock back and forth to the cadence of his steps down the tunnel. Slowly, the light disappeared into the darkness. Tommy, alone now, wondered what would happen if he made a mistake. With only one, two-hour long candle, courtesy of his insistent mam, ten hours of every day would be spent in the dark. On some days, the humidity would be so bad, that single candle would melt without being lit. And on most days the stench of human excretions mixed with the carbonic fumes of the bleeding coal seam was insufferable in the tight confines of the tunnel.

Chapter 4

⊱ • ⊰

TOMMY STEPPED OVER TO THE three-legged stool and sat down in the midst of the foul-smelling garbage heap. The soft flicker of candlelight cast shadows on the trapper door, the jagged walls tight on either side, the low ceiling just above his head. It was stuffy and humid. The silence was deafening. He took in a slow, deep breath, then he blew it out, hoping a calm might settle over him.

For the first time he heard a faint *drip, drip, drip*. From cracks in the rock wall, he saw the sheen of water. He reached over and touched the wet, glistening rock reflected in the flickering candlelight. He tasted it—salty. He listened closely, following its trickling trail to the floor, until it disappeared under the tracks.

His distraction from the smell of the filth was broken by the rustle of trash at his feet, then in the shadowed darkness to his left. He jerked up his feet to the rung of the stool, his heart pounding. He was not alone. *Twelve hours spent with rats is a long time,* Tommy thought. He blew out a pent-up breath through flared nostrils. *Six days a week is even longer.*

"I'm not afraid of you, so you rats better not try anything," Tommy shouted. He snapped his head toward a loud commotion in the pile of garbage as a rat seemed to scurry off at the sound of Tommy's voice.

For a long time, he held his tucked-up knees with trembling hands, concentrating hard on the rustling. "I can be brave, Papa," he whispered. "You just watch me."

After more than an hour, the rats seemed to grow comfortable with his presence and now rummaged all around him in the shadowed darkness. The rustling sounds filled his head. He struggled to inhale, to exhale. He whipped his head toward the sound of heavy rustling to see a much larger rat rummaging through the garbage beside him.

"You are my friends," he begged, struggling to make his voice sound confident. Dragging up from deep inside as much courage as he could muster, he shouted, "You . . . you are my new best friends!"

The scurrying stopped. All went quiet, at least for the moment.

Tommy tried to put aside thoughts of rats and concentrate on listening for a corve cart, readying himself to do his duty. He tried hard to feel for vibration on the tracks, but there was none. No sound but the rustle of garbage beneath his stool and trickle of water running down the tracks.

The minutes seemed to crawl by like molasses on a cold day as he prayed for someone to come. There was no reference to tell how much time had passed, what time of day it was. His mind wandered to thoughts of his mam, his little brothers, his sister. He missed Mam's stories. *I wonder what they're studying this morning? Maybe numbers?*

"I like those arithmetic games," he whispered, trying to distract his thoughts. "Especially when I win and get the bit of hard rock candy. Mam makes such good candy. I'm not sure I'd like a game anyway, if there's nothin' to win."

He continued to speak aloud to himself. "Maybe they're reading?" He remembered how Mam had taught him his alphabet. Carefully, she had written a big *T*, little *o*, little *m*, little *m*, little *y*. Then she had asked him to do the same. With his brows knitted together and his tongue stuck out in concentration, he had drawn out the letters. As he wrote, she whispered the sound of each letter in his ear. Taking care to enunciate each letter, he had repeated them.

"See there, you have written your first word. Now let's string the sounds together."

"T-o-m-m-y. That's my name. Tommy!" He remembered how the exhilaration overflowed its banks, knowing that forevermore he would be able to write his own name.

"If you work hard on the alphabet, you will be able to spell all kinds of words," Mam encouraged. "It'll take a while to learn the whole of it, but you can already read and write your own name."

He thought about the first sentence he had ever read. She had written it down in clear block letters, **You can do anything you put your mind to.** Together they sounded out each letter. Mam was patient. Slowly, he unraveled each word, then strung them all together into the sentence. He hadn't expected that words in a sentence could hold so much. When he strung sentences together into his first paragraph, his heart pounded, his hands sweated—he was hooked. And no one could stop him from reading. Not even Papa.

"I miss my mam." He tried to pull himself together. "But I'm a trapper now, makin' shillings for the family." Speaking out loud in the shadowed darkness made him feel less alone, so he continued. "Maybe I'll just blow out this candle and save it for later. Papa would be proud of me for thinkin' ahead."

He slowly spun around, memorizing the exact location of the door and trapper hole, the stool, the trash. He squeezed his brows together in concentration. "I'm not afraid of the dark. I can do it, Papa. You watch me." He took a deep breath, closed his eyes, and blew out the candle.

Almost immediately, he wished he hadn't done it. Opening his eyes in the pitch-black darkness of the mine's perpetual night, he could see absolutely nothing. His stomach leaped into his throat. He held his fingers apart just inches from his eyes—nothing. Only the memory of his hand etched into his mind's eye. The darkness was so oppressive; it filled his nostrils, his ears, and crushed in on his chest.

Tommy lifted his hands overhead and reached for the low ceiling. Taking shallow breaths through his nostrils, he blew them out through pursed lips, counting each one. In and out. He concentrated so hard he began to hyperventilate. He panicked. *I can't breathe!* He stood and hit his head on the ceiling.

"Ow." He held onto his head and rocked. Then, remembering Papa's admonition, *Ya'll be gettin' used to it,* Tommy forced himself to push back on his fear. "Be calm," he told himself aloud. "Don't panic."

Seconds seemed to pass like minutes, and minutes like hours. Tommy forced himself to concentrate on thinking of something other than the darkness and stuffy air. The smell of garbage seemed to intensify in the darkness. He fell to his knees beside the railway track. "I have to cover it," he said. "I have to deaden it."

Furiously, he threw fistfuls of dirt, coal tailings, gravel, and rock in the direction of the stool and the pile of smelly filth. The rats scurried away from the cascade of flying rock, dirt, and coal. His mind swirled in wild desperation as he scooped and shoveled faster until his sweaty hands were raw and bleeding. Sweat trickled down his forehead. His shirt stuck to his body.

Drained and emotionally exhausted, he slumped back against the wall, puffing hard. He wiped his filthy, blood-soaked hands on his dungarees. "I think the smell is better now," he muttered, trying to convince himself.

He wiped the sweat from his brow with his now-filthy shirt, then sat quiet in the darkness and let his mind wander. The noise of the rats returned to rustle in the garbage near his bare feet. He quickly put his back up against the wall, pulling his knees tight to his chest, wrapping his arms around them. Fear swept through him as he wondered if the rats would come gnawing on his toes.

Then, he remembered something his grandfather had told him. *Fear*, Grand had said, *is a natural feeling, nothing to be ashamed of. It's all in your mind, and ya can train your mind to make use of it. Just like ya can train yerself to shoot the bow.*

The bow, Tommy recalled, that was no easy task either. He still hadn't mastered it. Grand had worked the night shift in the mine, and every morning before breakfast, the two of them headed for Knabb Wood to collect plants, check traps, and hunt squirrels. An exceptional craftsman, Grand had made the only child-sized bow in the village. It was for Tommy to use for practice in the wood, strictly off-limits to villagers. Most dared not venture into it. And after all, there were snakes, rabid animals, and no real paths to follow, but there was also food if you knew how to find it. Grand taught Tommy invaluable lessons on identifying which plants were edible or medicinal, and which plants could kill. He taught Tommy about animals too—how to think like them, track them, hunt, and kill them. Grand's poaching from Knabb Wood had provided desperately needed food for their family.

"I miss you, Grand," he whispered, pushing back on the knot lodged in his throat.

"Face your fears," Grand had told him. "The awful ya know is less scary than the one ya can imagine."

There was a loud rustling in the pile of garbage. Tommy jerked toward the noise. Two beady eyes glowed faintly through the darkness at eye level. Tommy imagined them to belong to Papa Rat standing atop the pile of garbage.

Tommy sat up straight, staring at them for a long moment. "I'm not afraid of you," he spit out loudly in defiance.

The two beady eyes stayed fixed on Tommy.

From deep within, Tommy tapped into the courage to face his demons head on. "I'm bigger than you . . . probably. And you don't scare me . . ." Then whispered, "Well . . . maybe a little."

Tommy pulled back his shoulders in an effort to exert his will over the fearless rodent.

"From this time forth," Tommy declared fiercely, "your name will be Barnabas Bumble."

Confidence crept in as he thought only of fulfilling his promise to Mam. He would enunciate his words using his best King's English. "Would you like to join me in a spot of tea, Mr. Bumble?"

Tommy waited. The rat still didn't move. Then the eyes blinked.

Boldly Tommy added, "Would you like sugar or cream with your tea?"

Concentrating on speaking perfectly in his one-sided conversation with this apparent monarch of the varmint kingdom, Tommy forgot the rat's menacing eyes. In his mind he reshaped Mr. Bumble's intimidating demeanor into that of an imagined court jester. The idea brought an amusing smile to Tommy's face.

The human brain is the most miraculous instrument in the universe. Immersed entirely in darkness, it can create worlds from nothing, solely by the power of imagination. Tommy, in a world without a spark of light, filled it with the light of his imagination. His fears ebbed, the horrid smells lost their sharpness, and the pain from his scraped-up hands, knees, and bumped head were forgotten. At least for the moment.

Of course, he presumed it might be a long time before Mr. Bumble could carry on a decent conversation. "But if it's time I need, I suppose I'll have plenty of it."

Tommy felt the rumble of the tracks beneath him. Startled, he looked around in the darkness. Then he put his ear down to the rail and felt the vibrations.

"Someone's coming!"

Tommy's heart leaped back into his throat as he jumped to his feet to man his post. Whacking his head on the low ceiling, he fell back, rolling in the garbage. Muttering, he got back up slowly this time, rubbing out the ache with one hand and sliding the other hand along the door until he reached into the trapper hole. He grabbed the pull rope inside, put his foot on the wall, and with both hands, leaned away from the door, pulling with all his might. The heavy wooden door squealed open, revealing the illumination from two oil lamps bobbing up and down in the distance. The comforting light began to fill the tunnel. Relief flooded his heart. He decided right then and there that he would never again take the security of the light for granted.

As the sound of metal wheels on iron track grew closer, the shadowed outline of the coal cart and the two boys pulling it up the track came into view. Tommy wedged the doorstop stone against the base of the door to hold it open well ahead of his first customers, then stood waiting impatiently for their arrival. He smoothed down his shirt, tucking it into his trousers, and slid his filthy hands through his hair in the hope that he might look a bit more presentable.

The hurrier at the front of the cart had a harness wrapped over his shoulder. He leaned almost parallel to the track, pulling the empty coal cart up the gradual incline. Grunting with all his might, he placed one foot in front of the other on the crossbeams, his muscled legs straining as he slowly pulled the cart up the grade. The thruster pushed from behind with hands and head pressing up against the cart's iron back plate.

"Hello!" Tommy called out well before their arrival, thankful to be speaking to someone besides rats. "Door's open. You don't have to worry!"

As the two boys drew closer, Tommy rushed forward and fell in beside the thruster, helping him with the final push up the grade to the flat track where his trapper door stood propped open.

"May I beg a light off your lamp for my candle, please?" Tommy politely asked as the cart pulled to a stop inside the doorway.

The thruster turned with a smile to look at Tommy, sweat dripping down his face. His confused brows pulled together in a frown, then raised in curiosity. He swiped his dirty hands down his dungarees, nodded, and added a grunt and a giggle to complete his greeting ritual without a word. He was a skinny, freckled-face boy with the biggest ears Tommy had ever seen.

"Thanks for the help, mate," the boy finally stuttered out in answer. "Ya speaks funny."

Before Tommy could reply, the hurrier in front called back to the thruster. "We be a-pullin' up here, Fidget!"

Fidget pulled a wooden sprig from his belt and jammed it under the wheel to keep the cart from rolling back down the grade, then stepped away from the cart.

The hurrier in front, who looked to be a couple years older than Tommy, unstrapped the gurl belt—a leather strap for pulling the cart, wrapped diagonally across his chest—and walked back to stand with them. Without speaking further, he pulled off his lamp, took out his candle, and held out the flame. Tommy reached out, lit his candle wick, and the boy put the candle back into his lamp.

"Thank you."

"This here be your first day as a trapper, heh?" said the older boy confidently, clearly the one in charge. Tall, strong, calm, and more serious than his deferential thruster, he seemed much older and self-assured than his years would suggest.

"Yes, sir. I'm Tommy Wright. Your name is?"

The hurrier seemed surprised at being addressed so formally, but still he answered the question undisturbed. "My name be Tutor Turton. And this here be me little brother, Fidget."

Fidget acknowledged Tommy with a tip of his cap and a twitching smile.

"How comes ya talks funny?" Fidget asked, jiggling his loose-fitting dungarees up and down with a hop.

"My mam be teachin' me to speak proper English," Tommy self-consciously declared.

Fidget grinned. "I like that! I be wishin' me mam would teach me to talk like that." He wiped his hands on his pants and tweaked his ear.

"How long have you been a hurrier?" Tommy asked Tutor. "I'm hopin' I might become a thruster in a few months and a hurrier when I'm nine."

"Well, you'd be a sight quicker'n me." Tutor frowned, brows pulled togehter. "Took me a year afore bein' a thruster and just got the job as hurrier when I turned ten. Fidget here is almost nine. But whatta ya wanna be a hurrier for anyway? I ain't seein' the draw. And a thruster . . . Fidget, show 'im your head."

Fidget giggled, pulled off his cap, fiddling with it from hand to hand, then lowered his head.

Tommy's eyes widened when he saw the one-inch-high bump and bald spot that had formed where Fidget pushed the cart with his head. "Wow, that's really somethin'. How long does it take to get it?"

Fidget bobbed up and down as he chuckled. "Lot sooner than to me likin'," he said with a twinkle in his dark brown eyes, tweaking his coal-blackened nose.

"Come 'ere, see what it's like to be a hurrier." Tutor pulled up his shirt and Tommy could see a wide callused band stretching diagonally across his muscled chest from his shoulder to his waist.

"It's where the gurl belt rubs." Tutor traced the lines of the wide welt across his chest with his finger. "The first day, ya get blisters as big as toads. The next day, the strap pops 'em and they stick to your shirt and trousers. Then ya get new blisters, and it starts all over again 'til the skin turns to leather and gets thicker and thicker, 'til it looks like this here leather whippin' strap. And take a look at these bunged up hands and knees!" He pointed to the raised knots on his knees.

"Wow!" Tommy exclaimed in admiration. "Looks like the soles of me bare feet!"

"Ya can't even eat your food with your hands 'cause they's so sore and mangled for months."

Tommy reverently inspected Tutor's scars, feeling privileged to touch such an intriguing badge of honor. "That's really somethin'." He pulled his shoulders back. "I can do it. I can be a thruster, and a hurrier too."

"Well, good on ya, mate," Tutor responded. His white eyes shined bright out of his blackened face, and his wide, contagious smile revealed crooked, blackened teeth.

With that, they'd begun the process of becoming journeymen in the lifelong brotherhood of miners.

Tommy stood silent, gazing at Fidget and Tutor Turton pushing the cart up the track, their bobbing oil lamps disappearing into the darkness. If all went well, sure as warts on a frog, it would be no time before their brief acquaintance grew into an abiding friendship in the way boys do best.

⤚ • ⤙

Tommy came home that first night exhausted. Papa had gone to the pub, but Tommy hardly had enough energy to wash up and eat the dinner Martha had waiting for him. After he had fielded questions from his siblings of the day's events during dinner, Mam sent Joey and Edie outside to play in the cool of the evening. Georgie stayed in on Tommy's lap, for he would not be separated from his favorite brother.

"So, how was it down in the mine with Papa?" Martha probed gently.

"Ya shoulda seen him, Mam!"

"Proper English, Thomas."

"Sorry. Papa walks so fast! I could hardly keep up with him. And he can crawl fast, too!"

"Really?" Martha had a pensive look in her eyes as she continued washing dishes, one of her never-ending chores that prevented her hands from ever standing idle. "Did you have any difficulties?"

"I knew I would be opening and closing trapper doors, but now I know why. It's really an important job." Tommy furrowed his brow. "I was a little worried I might not do it right." Tommy looked down at his feet, remembering the many fears that had coursed through him that day. "But I wasn't scared or anything." He looked away from Martha's penetrating eyes. "Really, Mam, I wasn't." He looked down again and whispered, "Well, maybe just a little."

Martha picked up Tommy's dish, washed it, and put it on the counter, giving her son the time to share more without pressing him. After four generations of coal miners in her family, watching her own brothers go through similar experiences, she knew most of what went on in the mine.

"I was a little scared to get into the cage, but not too much. And maybe a little in the mine too . . . but you know I'm not afraid of the dark." He frowned. "There's rats down there, Mam," he whispered, wanting to tell her more, but when he saw her shining eyes looking down at her hands drying dishes, he looked away and tried to avoid meeting her gaze. When he glanced back sheepishly, he found her eyes searching his face.

"Don't you worry about me, Mam," he declared, attempting to convince her with his most confident voice. "I'll be fine."

"It's okay if you're scared, Tommy. It was your first day, after all. I am sure I would have been scared."

"I didn't want Papa to know I was afraid," Tommy confessed in an almost inaudible voice, his lips curling down. "What if he told me I couldn't come to the mine with him anymore? I mean, what if I couldn't be like the other boys? What if I couldn't become a man like Papa and Grand?" Tommy sat silent, staring into the fire and holding a sleeping Georgie on his lap. Though he wanted to tell her everything, he just couldn't. He knew she would only worry about him, and he couldn't bear to see her cry again.

Martha wiped her wet hands on her apron. Then, with eyes rimmed in scarlet, cheeks puffy and pink, she knelt down in front of her little boy. Putting her hands on his knees, she looked deep into his eyes. "I'm proud of you, Thomas, and I'm sure Papa is too. It's okay to be afraid. Courage is learning to not let fear hold you prisoner. You just do what needs to be done."

Out in nature fawns were playing in the woods, young birds rustling in their nests, and in the big houses little children pressed against their mother's breast for comfort. But in the coal mine, fearful little boys and little girls faced an unwelcoming world of darkness, devoid of tenderness, where they were left in tears. With the innocence of youth left behind, these pale, discouraged, hopeless children would drag their burdens through dark underground tunnels, danger at every turn and fear in every foreboding corner. They were destined for early graves. For some, they looked forward to a heaven with no coal mines, where fear was no more, imagining their days filled with picking flowers in the sunshine.

ᐳ • ᐸ

Tommy's first payday came at the end of a long week. Proud to now be a contributor to the family income, he gave Mam the whole ten pence, but she gave him back two and closed his hand over the four half-pennies. "This is for you to do with as you wish."

That Sunday after church, Tommy, good to his word, gathered up his little brothers and sister and headed into the village square. With Georgie's hand in his, Joey and Edie skipping along behind, he took them to the Silkstone Company Shop that sold everything from expensive picks and shovels, candles, and lamps to coal candy.

"Tommy, where we goin'?" Joey asked.

"We're on a coal candy mission," Tommy advised.

Georgie threw his arms around his big brother's waist in delight. Coal candy was a delicacy made from sweet licorice root commissioned by Lord Fitzwilliam to celebrate his coal mines. It came in any color you wanted, as long as it was black.

"No. Really?" Joey responded.

"Me, too?" asked Edie, making sure she wasn't forgotten.

"You too, Edie."

"Yer the best brother ever."

The chime rang out as Tommy opened the door to the Silkstone Company Shop. He stepped up to the counter. His little brothers and sister stood behind him. Their eyes were lit up in childish anticipation, mouths hanging open in awe, imaginations running wild, as they eyed the glass jars of delicacies on the counter.

"I'd like to buy my brothers and sister here each a piece of coal candy," he said to the shop's clerk behind the counter, who was looking down at him. She wore on her face an ominous frown of displeasure.

"Well, how much ya got?"

"I got two pence! My first pay from workin' the mine."

She looked down at Tommy, then to Georgie and Joey, whose dirty hands and runny noses pressed tight up against the counter's edge she had just cleaned. Their eyes were fixed longingly on a particularly enticing striped curlicue piece in one of the glass jars on the counter. Edie patiently stood behind with her arms folded shyly looking up at the lady, her cherub face shining in anticipation. They had coveted the coal candy in those jars on many an occasion during Mam's trips to the shop.

"That stripy one there in the jar wouldn't be a penny by chance, would it?" Tommy, with hopeful eyes, pointed at the jar.

The shop's clerk placed her hands atop the counter, leaned over, and looked down long and hard at each of the four children, who were staring at the exquisite, particularly beguiling piece of coal candy.

She hesitated. "Yer buying' it for your brothers and sister, are ya?" she asked.

"Yes ma'am."

Her eyes softened as the four wide-eyed children looked up at her longingly.

She looked at the four half-pennies on the counter. "The curlicue delectable?" she asked, looking at the licorice. "No, son. Them's be a halfpenny each."

With a proud smile spreading wide across his face, Tommy's eyes lit up. He pushed the four copper half-pennies toward her.

"We'll take four of 'em, please!" he exclaimed in a proud moment of exultation.

She reached in the jar and picked up the four pieces of coal candy, and handed one to each child. For fear of her changing her mind, the children said their thank-yous and scampered for the door, each tightly gripping their prize as if it were the most precious possession in their lives. As they ran out the door, they giggled and jumped up and down like nobody was watching.

>• ◄

After they left and the door had closed behind them, a miner who had watched the whole affair from the corner of the shop looked up at the shop's clerk behind the counter. "Now, Josie! You know those licorices is a penny each."

"Well, what's it to ya, McConnell?" she huffed, wiping the snot and dirty fingerprints off the counter. "Today they's on sale."

He sauntered up to the counter with his purchase, putting his hand deep into his pocket, and frowned. "Best not be tellin' Sir Robert. He ain't never gave no sale in his life." Pausing for a moment, he added, "Good on ya, Josie. That boy's got a hard life ahead of 'im. He's gonna be needin' a few good turns." He put ten pence down on the counter.

She took his money and went to the cash box, but when she turned back around to give him his change, he was already at the door.

"Hey there, McConnell. Ya forgot your change!"

"Keep it." He smiled back. "Ya knows, Josie, ya ain't as hard as ya let on."

Chapter 5

≻ · ≺

A FTER FOUR MONTHS OF WORKING in a remote chute of the mine, Tommy completed his probationary posting and was reassigned to a busier post in a larger drift tunnel. His apprenticeship had served him well, and he now considered himself a veteran trapper. At this new posting, time would pass more quickly with the hustle and bustle of added traffic, but here the ventilation and flushing out of poisonous gases through the proper operation of trapper doors was even more important. Many young miners' lives depended upon Tommy's conscientious attention to his work.

At Tommy's new post, an almost-grown boy could stand to his full height. Lanterns were hung every fifty feet, and shed enough light so that he could read when he wasn't taking care of business. This new assignment was not nearly as stiflingly warm and humid, nor wet and muddy. It had a shelf for his books, and a two-holer loo just down the tunnel. With all the comforts of home, Tommy could now keep his post clean and organized.

Most days, coal carts trundled to and fro on a regular basis. Pulled and pushed by both boys and girls, the carts were often stacked or linked together in three or more cars, sometimes with more than one hurrier and thruster working in tandem.

Best of all, there was a fellow trapper stationed just a few hundred feet up track.

"Hello!" Tommy called up the drift shortly after arriving at his new posting, eager to make introductions.

He waited. There was no answer.

"My name's Tommy Wright!" Again, he waited, but no answer.

Just then, a boy emerged from the shadows, sauntering slow and easy toward Tommy's station. Maybe a year older than Tommy, he had freckles covering the entirety of his impish face and flaming red hair that stuck out in all directions. With a pug nose, big ears, and bright green eyes that sparkled with intelligence, he possessed a smile that seemed to suggest there was something he knew that others didn't.

"I decided to wander down the tunnel to see what all the ruckus was about," announced the boy. His face widened into a grin that revealed two blackened front teeth Tommy imagined had come from some errant fall. "I brought some of me best stuff over." The boy reached deep into one of the pockets of his homespun bib dungarees. Sewed specially by his mam, they seemed to be four sizes too large and have pockets everywhere. The cuffs had been rolled up above his ankles, exposing blackened, calloused bare feet.

"This here's Stick Man. I made him. He's a Trojan warrior!" The boy handed over the mangled branch carved into a stick figure of a man. "I brung 'im as a 'welcome to the neighborhood' present!"

"Thank you."

The boy reached deep into another pocket and pulled out an old used snuff box. "You'll be needin' this!" he asserted. "It's chock-full of clipped fingernails and a possum foot."

Uncertain what to think, Tommy looked up at the boy, who reassured, "It's a surefire way to ward off snakes."

Tommy's eyes were wide with interest. "I haven't seen any snakes around here."

"See there," the boy grinned. "What did I tell ya? It works every time."

Tommy couldn't argue with that logic. "Well, good then," he replied, excited to have a new friend.

"My name's Tigger," he declared. "I'm hearin' yers be Tommy."

Tommy nodded. "I'm pleased to meet you, Tigger!" he said, using his best King's English.

Tigger whistled. "Whew! Mr. Prudent told me 'bout yer talkin', but that's some mighty fine speakin', Tommy me boy!"

"Who's Mr. Prudent?"

"Tutor Turton. I call 'im Mr. Prudent. Him bein' so sensible and all."

"Tigger, is that really your name?"

"Well, me mam calls me Mr. John Gothard when I'm about to get a whoopin', but otherwise, everyone calls me Tigger." He let out a mischievous chuckle that ended with a little snort from his freckled face. It was laughter Tommy was sure could melt cold hard steel. He couldn't help but giggle right along in delight.

Tigger paused a moment to consider what was on Tommy's lap. "That's a mighty fine rat ya got there."

"Oh, this here's my friend Barnabas Bumble." Tommy passed a comforting hand along the big rat's back. "I call him Barny when we're gettin' on fine and he's not wandering off in one of his curious ways."

"Can he roll over for food?"

"I suppose he might with some coaxing. He's mighty smart."

Tigger looked Barny over. He hefted him in his hand and looked into his eyes. "I'll have to bring by me Harvey Rat. What's them books ya got there?" Tigger asked, pointing to Tommy's stash on the shelf carved into the stone wall.

"Why, I've got *Gulliver's Travels, Robinson Crusoe, Aesop's Fables.* This here is my writing journal, and . . . !"

"Whew!" Tigger exclaimed, forcing his eyebrows together into a frown, then raising them into a wide-eyed expression of hope. "Got any venturin' in 'em?"

Tommy smiled. "They do!" He held out *Gulliver's Travels* to Tigger, who took it, held it, and looked it over. "It's about a fellow who goes exploring."

Tigger's face brightened, but then he handed the book back. Glancing sheepishly down at his feet, he said, "I cain't read." When he looked back up he asked, "Can ya read me some?"

"All right," Tommy promised.

Tigger sat down on the stone floor, crossed his legs, and put his elbows on his knees. "How 'bout now?"

"Now?" Tommy questioned with a giggle, then shrugged his shoulders. "Okay—I suppose I could! But don't you have to be at your trapper post?"

Tigger's trapper post was different than most because of the steep grade leading down from the coalface to his posting. It required the hurrier to actually hold back the cart coming down the steep slide toward the door.

"I got plenty a time to get back after the shakin' starts."

"Well, all right then."

Tigger curled his hands up under his chin and, with rapt attention, prepared to be entertained.

Tommy smiled. There was no question about it, he was going to like this fellow. He opened the book and began to read. "'My father had a small estate in Nottinghamshire; I was the third of five sons . . .'"

Tigger was enthralled by the story of Gulliver. His enthusiasm was so catching that Tommy couldn't help but be engaged in the telling.

"That's somethin', Tommy me boy," he gushed, enraptured by the first chapter of the story. "Can ya read me more sometime?"

"Okay."

"I sure do like that Gulliver fella!"

"Maybe, if you want, I could teach you to read?"

"That would be somethin'." Tigger looked on with a clear eagerness to learn. "Mam says I just might be somebody someday, unless o'course they hang me first." His laughter shook loose from his expressive, freckled face. "Someday, I want to be an explorer and visit places like this here Gulliver fellow. Climbin' mountains, trackin' lions!" Tigger rushed on, thoughts coming on faster than he could stutter them out.

⤙ • ⤚

There comes a time in every boy's life when he has a raging desire to hunt for adventure, dig for buried treasure, and make guns out of sticks to shoot the evil bad guys. Tigger was no different. He just had a bad case of this boyhood disease. He always seemed to be conjuring up some kind of earthshaking scheme or daring escapade.

As the weeks rolled on, Tommy came to look forward each morning to the sight of Tigger sauntering down the tunnel toward Tommy's trapper post with a mischievous gleam in his eyes. The closer he got, the more Tommy felt the muscles in his face relax, his ears perk up, and the anticipating smile ease across his face. He had found someone he could be himself with, someone to share his thoughts and the mysteries of boyhood. Tigger's enthusiasm for life connected with Tommy. As the days rolled on, the warmth of friendship simmered into an unshakable affection that rose from somewhere deep in Tommy's soul.

Over the long autumn weeks, the two boys became almost inseparable. Tommy gained a certain reverence for Tigger's imagination. Sometimes, it scared the holy bejesus out of him, but overall, his enthusiasm was downright inspiring. And they seemed to finish every excursion into their imaginary world with contagious laughter. Sometimes, it seemed planning for their adventure might be better than actually finding it.

One day, Tigger wandered down to Tommy's post.

"Mornin' to ya," Tigger said. "I be rememberin' your brother Joey's got the chicken pox, so I brung ya a poultice."

Tommy grinned. "That's thoughtful of you."

Tigger seemed to have an arsenal of home-brewed remedies to cure just about any ailment a body might have. "Ya mixes these cockroaches and leeches up real good with spittin' juice and put it on your chest in the night, and them pox won't touch ya. Oh, and don't forget to put your shoes cattywampus at the foot of your bed, for connectin' with the vibrations."

"Sounds fascinating. Really works, huh?"

"Ya doubtin' me are ya? Of course it works."

"One thing I know for certain, I ain't got no shoes," Tommy lamented.

"Yeah . . . that's a problem we all gots," Tigger confirmed with a giggle. "Ya tried the wart remover tonic yet?"

"I was planning on puttin' my hands in that sack of snail guts tonight. Dang Barny rat keeps peein' on 'em."

Tigger nodded. "Best do it soon, 'cause good snails is hard to come by."

Tommy's mam didn't send him to work with much for lunch, but Tigger had none at all. With his papa laid up in bed with black lung, it took all the money Tigger earned just to pay for the poultices, herbs, treatments, and the barest of necessities for their family. Still, Tigger was too proud to ask Tommy for food without just compensation. So, when an opportunity came for Tommy to share a bit of his lunch with his best friend, he did some trading.

One morning, Tigger showed up with a sly smile on his face. "See what I brung ya?" Tigger reached into one of his endless pockets and pulled out a toad. "This here's Croaker Toad. I been a trainin' him to jump. It takes a heap of coaxin' and charmin' to get a proper jump outta him. But I swear he can jump clean across me trapper post!" Tigger puffed out his chest in unabashed pride.

"Boy, what I'd give to see that!" said Tommy, fascinated.

Tigger's eyes twinkled with expectation as he set the hook and reeled in Tommy. "Well, I don't know, Tommy," Tigger opened the negotiations. "It hardly seems right puttin' 'im on the spot after workin' hard all mornin'. He's plumb tuckered out. I hate to be takin' yer lunch."

"I'd give ya my muskrat stew just to see him try it."

"Well, maybe I could set him down for a spell . . . but I cain't promise nothin'."

"Fair enough." And the deal was struck.

"Ain't no guaranteein'. He's mighty temperamental," warned a crafty Tigger as he set Croaker Toad down and began to coax him on for a demonstration. But no matter what Tigger did to encourage ole Croaker along, he refused to move a muscle. And after a time, Tigger stepped back. "Looks to me like he's plumb tuckered out." He shrugged his shoulders, frowned, his eyes mischievous as ever.

Tigger gathered up his toad and stuffed him back in his pocket.

On his way back to his post, soup in hand, he called back to Tommy. "I'll bring ole Croaker by tomorrow when he's a feelin' a bit peppier. Don't worry, ya'll see some jumpin' like ya ain't never seen afore."

Tommy knew he'd been conned, but he was happy to let Tigger think he'd earned it fair and square.

＞・＜

One morning Tigger showed up so excited, he couldn't wait to share the news on his latest invention. "I call it me sound-raising machine. Ya see, Tommy me boy, I've connected two tin cans together with a long, waxed string." Tigger's eyes danced in his freckled face. "If we stretch it real tight, we can talk to each other without no yellin'."

"Really?" Tommy was hooked from the start. "Show me."

Always eager to try Tigger's inventions, Tommy took one of the tin cans and held it to his ear. Tigger walked toward his own post, stretching the waxed string well outside the normal range of hearing, then spoke into his tin can. The vibration carrying Tigger's voice traveled along the taut line, and as Tommy listened, his eyes widened. "I can hear you!" A surprised smile spread wide across Tommy's face. "Tigger, it works!"

"O'course it does." Tigger's brows came together into a pronounced frown, perplexed at why Tommy would even doubt the quality of his inventive magic for a moment. "And iffen I can find enough string, I'm gonna reach all the way from my trapper post to yers."

Tommy stood in awe. "You're a regular genius."

"That I am, Tommy me boy!" Tigger responded, chuckling.

Tigger was the kind of boy who could find fascination in anything. He was loaded with questions about everything. But the way he figured it, an education might just get in the way of his imagination. To him, the whole world was a wonder, and his ingenuity seemed his sole link to a sane world during those long, mundane hours leashed to his trapper stool. He was insightful, inquisitive, and possessed an indomitable will to make implausible plans for a life as an explorer, an adventurer, an inventor. He inspired Tommy to explore and use his own imagination—in ways he had never before considered.

Chapter 6

⋗ • ⋖

O VER THE REST OF FALL and into winter, Tommy, Tigger, and the Turton brothers wrangled as much time together as possible during their long work weeks of twelve-hour shifts. Frequently, Tutor would have Fidget put a twig under the wheel of their cart at Tommy's trapper post and the four of them would talk, laugh, and do their level best to one-up each other, as boys do in most all things. They were developing friendships that would last a lifetime. Sometimes, they would practice the time-honored tradition of trying to outdo each other with insults. It was an art, really, and each boy hoped to earn the cherished title of "King of Insults."

One afternoon, Fidget and Tutor started in on each other.

"Yer a moron," Tutor ridiculed Fidget.

"Well, yer a scab-eater, Tutor!" Fidget returned.

"You lick toe jam for breakfast!"

"If my dog was as ugly as you, I'd shave his butt and have him walk backwards!"

"Well, you bob for apples in that two-holer over there!"

Fidget flared his nostrils. "Yeah? Well, that ain't nothin'. You pull the corve like a girl, Tutor!"

⋗ 75 ⋖

"Yeah? Well!" Tutor sputtered, searching his bare toes for a response, but found no words there. Mr. Prudent was left speechless in defeat. What could a boy do to answer a slur like that? They had sworn their sacred blood oath to have nothing to do with those insufferable creatures.

Fidget Turton folded his arms in triumph, reveling in the limelight of victory, the King of Insults, while Tutor hung his head low in abject humiliation, knowing he'd been bested.

⇒ • ⇐

Through the long cold winter months and spring thaw, Tommy and the rest of the children in the mine seldom saw the sun. It took a toll on Tommy but was merciless on Tigger. Now, with the start of the summer months, Tigger was in a terrible funk, longing to get out into the sunshine. One morning, when he didn't show up at Tommy's post for his venturin' reading, Tommy was concerned. Tigger never missed a reading day. Tommy now had a whole host of books—*Swiss Family Robinson, Gulliver's Travels, Cast Aways, Milton's Science Primer*. Mam seemed to find books for him in the most unlikely of places.

"Tigger?" Tommy called out, only to hear silence. "Tigger!" Tommy's voice echoed back unanswered. He tried rousing him through the sound-raising invention. Still, no answer.

Worried for his friend, who'd become more withdrawn and despondent as the winter wore on, Tommy decided to wander down to Tigger's post. He found his friend sitting on his stool, his head in his hands, staring at his feet.

Tommy sat down next to him. "You alright, Tigger?"

Tigger sighed. "There's got to be more to life than openin' and closin' trapper doors," he moaned. "I cain't do it anymore. Ain't no part of me future plans, and I just cain't."

"Aww, Tigger. It's not so bad, is it?" Tommy put a comforting hand on Tigger's shoulder. "Here, take my honeyberry jam sandwich."

"Naw, your mam made it fer ye."

"Hey, it's not like you to turn down a sandwich. Take it." Tommy held it out.

Tigger, having been enticed beyond the limits a boy could reasonably be expected to resist, wiped his runny nose with the back of his sleeve. He reached out and took the sandwich with a thank-you.

"Thank you. Suppose I cain't pass up the best honeyberry in shantytown." Tigger munched. "I gotta get on with venturin' or my life is as good as a snuffed-out candle."

Tommy tried to be encouraging. "Well, we got each other, don't we? You're my best friend for life, Tigger!"

"I s'pose that's true, but I'm not likin' bein' told what to do. I'm decided. Iffen the company be tellin' me there's something I cain't do, well, from now on I'm all about it," Tigger told Tommy in downright defiance. "If they be tellin' me I'm obliged to do it, I ain't doin' it no more."

"What kinda talk is that, Tigger? We're trappers. And someday, we'll be coal miners just like our papas."

Tigger shook his head. "I want to explore dungeons and castles and Indian villages like this here fella Gulliver! Not spend my days waitin' for a train full of coal to come by." Tigger was silent for a long moment. "We need to get to venturin' if it's gonna be our callin' in life."

Tommy sighed. "Look, Tigger, our families are coal miners, been that way for a hundred years. Face it, we're gonna be coal miners too—til the day we die. Heck, I'll probably be blown to bits like Grand, buried alive forever in the pit."

Memories of the day he lost Grand swarmed into Tommy's mind. The news had come on a warm sunny morning in May. He'd just finished dumping a gunny sack of coal he had found into the bin beside the hearth when there was a knock at the door. Tommy could still smell the sweetness of springtime mixed with the heavy, bitter taste of coal dust on the easterly breeze when he heard the man at the door.

"There's been an explosion down in the mine," the man had told Mam. Then he said Grand had been incinerated in that awful pit.

The impact of the messenger's words had knocked the breath right out of Tommy. Light-headed and dizzy, he had doubled over, slumping to the floor in disbelief. His stomach seemed to turn inside out, and there was a tightness in his chest. He was unable to move, unable to speak, unable to comprehend the meaning of the words. Racking sobs tore at his heart, almost ripping it right out of his chest.

After the pain came the unquenchable anger. How could Grand let this happen? Why had he left him alone? Looking toward heaven, Tommy had pleaded with God to tell him why, but there was no one up there who could answer his pain. It had been two years but still, it was difficult to suppress that hurt.

Tommy reached deep into his pocket and held tight to the lucky rabbit's foot. *I miss you, Grand*, he thought.

". . . Tommy? Ya alright, Tommy?"

"I'm sorry, Tigger. But I suppose what I'm sayin'," Tommy blurted out, "is the chances of either of us ever gettin' outta the mines are about a million to one."

"Really?" There was a long silence. "Them's better odds than I was thinkin'." Tigger's brow came together. He took in a deep breath, stood up, and let it out. His frown turned upside down.

Tommy rolled his eyes. "Aww, Tigger, whatta ya thinkin'?"

"We best get crackin' on the venturin' trail."

"Oh, Tigger." Tommy threw his hands in the air in resignation, then paused for a long minute. "Well . . . I've been ponderin' on a new game," he continued cautiously. "It could be a pretty exciting game too. But I haven't wanted to mention it 'cause it could also be dangerous." He paused. "I'm afraid you're not trustworthy enough to be careful."

Tigger's ears perked up, his brows raised. "Dangerous? What's wrong wit' a bit a danger?"

"You don't give up, do ya?" Tommy sighed with confirmed misgivings. "We could get ourselves in a lot of trouble—or worse, maybe even bring the Candyman down on us."

"Whatta ya know 'bout the Candyman?"

"I seen 'im once."

"Where?"

With a shudder, Tommy recalled the day the brutal company enforcer, Candyman Boo Black, had come to their shanty to shut down Mam's school room. She'd been in the middle of teaching some of the neighbor boys their ABCs.

"Without knocking, Boo pushed open our door bellowing, 'This shanty ain't fer no school. Them's the rules!' Mam pushed me and the rest of her students behind her. She planted her feet wide apart, like this"—he showed Tigger the stance—"and dared the scary monster to knock her down."

"No, she didn't." Tigger's dancing eyes were wide with admiration.

"She did." Tommy recalled holding onto his Mam's skirts from behind, peering around her waist to look up at the towering, brutish beast of a man who sneered down on him. Tommy's heart was pounding in his throat, fear gripping every part of him. "Boo's only got one eye." Tommy could still see that livid, bloodshot cyclops eye, glaring at him, the sweaty bald head and thick eyebrows, black and arched like the wings of a vulture. "Pocks on his face, he smelled like a whisky still. His big scar ran across his forehead to an eyepatch over his bad eye. It gives me the chills to even think on him."

Tigger hopped up and down. "No way!"

"'These little nits ain't got no need fer readin' fer diggin' coal,' Boo said, spitting out his words in his awful putrid breath," Tommy recalled. "He had yellow tobacco slime drippin' from the corners of his mouth. And two silver front teeth." Tommy remembered them shining, amidst a graveyard of his remaining rotting teeth.

"No way. He didn't say that?" Tigger shuddered. "That man would scare the bejesus outta me."

"He did!" Tommy remembered watching the cyclops search his face for fear. "It turned my stomach inside out. But Mam was amazing. She had no fear. She just stared into Boo's evil eye under those squirmin' caterpillar brows."

"Did yer mam say anything back ta him?"

"I'll never forget it. Grand used to say she seemed calm as a summer's day under fire. She stared right into that cyclops eye. 'Rules are not to be followed,' said she, 'if they go against the laws of God.'" Tommy paused, remembering the moment. "'The company may own everything around me,' she said. 'They may rule my life, even control my body with an iron fist. But my heart, mind, and soul are mine! You will not stop me from teaching these children to read. Do your worst, if you will?'"

"Oh my gosh. Did he try to kill 'er?"

"He didn't do nothin'." Tommy remembered how the maniacal company enforcer just stared down at Mam. "'I'm waiting,' said she. 'Fer what?' said he." Tommy paused, in awe of his mam, remembering how he'd hid behind her skirts the whole time. "'I'm waiting for you to leave my home . . . or death . . . whichever comes first.'"

"No, she didn't say that." Tigger's wide dancing eyes were filled with respect for Tommy's mam. "Ta tempt fate like that."

"She did."

"What did Boo do."

"Nothin'! He just stood there giving each one of us his terrifying look." Tommy stared off into the distance, remembering. "Then he turned and walked out."

"Amazin'!" Tigger stood in awe, his mouth hanging agape.

Tommy remembered being plagued with nightmares for weeks after. He'd woken up sweating, his undershirt drenched, screaming once, that roving eye after him. "None of the neighbor kids ever came back to Mam's school. Their mams was too scared to let them."

"Yer lucky yer mam didn't get kilt."

"He hit my papa once too." Tommy recalled that birthday night. "If I ever see him again, I'm gonna stand up to him for what he done to my mam and papa."

"Ya knows how he got that name, Candyman Boo?"

"No . . . how?"

"His pap was an ole drunk—beat Boo somethin' awful. It made him mean from the start."

"Yeah?"

Tigger went on. "Boo was a bully. Bigger than everyone else. Me pa tells me he scared the bejesus out of the other boys."

"I've seen boys like that scaring other boys for the sheer pleasure of watching 'em squirm in terror."

"That was him," Tigger continued. "They say one day, he lured a little girl in with rock candy. She was holdin' her kitty-cat in her arms."

"Oh no."

"Yup, Boo took the kitty from her, tied its feet together with a slipknot, doused the cat with lamp oil, then made the little girl watch as he lit the cat on fire. The poor thing ran in circles, screeching, flames trailing, until . . . All the while the little girl's screamin' her guts out."

"Oh . . . my . . . gosh. And that's why they call him Candyman Boo?"

Tigger looked off into the distance. "The name stuck. And he done worse than that to me pap!" Tigger shook off the thought. "'Nough of this kinda talk, Tommy me boy. Let's get to venturin'.

"I don't know, Tigger."

"Courage, me boy. We can play up your way, iffen yer afeared."

Tommy knew full well that Tigger would hound him mercilessly, and he wanted to pull his best friend out of his doldrums. This new game might just do it. So, with a bit of uneasiness, he began, "I call it Chutes & Daggers, a game of vision and cunning that is sure to refine yer venturin' skills."

And so it began. Although it was not so unusual for young boys to play exploring games during their shifts, it was, of course, strictly forbidden for them to leave their trapper posts unattended for any length of time.

With infectious determination and an impish thirst for adventure, Tigger embraced the game with reckless abandon. Soon, both boys were all in, playing their elaborate brand of hide-and-seek with a gusto seldom seen in a boy. They plotted, strategized, schemed, and stuck to each other like cockleburs on woolen socks.

On the first day Tigger won the match, and the second day as well. Like most boys, he enjoyed outdoing his best buddy, and stretched the truth of his victories into a saga of conquest.

"I'm thinkin' I'm faster, stronger, and wilier than you," Tigger advised.

Tommy scoffed. "You think so, huh? I wouldn't count your chickens before they hatch."

"Ya best keep yer wits about ya, 'cause I know all about me chickens. It's just a matter o' time afore ya slink away like a polecat fallin' down the hole of defeat."

At first, both boys tried to listen for oncoming traffic and remained in tune with the rumbling vibration of the tracks. But as the competition rolled on, day after day, the rivalry between the boys intensified, causing them to put aside any caution they might have otherwise had. And Tigger didn't have much to begin with.

"I'll be takin' ya to the woodshed on this one, Tommy me boy," Tigger bragged, after holding onto a slim lead in the competition. It was, of course, in Tigger's nature to take the biggest risks and embellish the tallest tales, sometimes even scaring the bejesus out of himself, if it meant trouncing the living daylights out of his best friend. Nevertheless, Tommy quickly caught up, winning most of the following weeks of competition to tie it up. Every day, the competition grew even more intense, each fiercely pursuing the win and the crown of thorns.

Tigger dodged in and out of the shadows, confident he was on the road to breaking the tie. Lost in the enthusiasm of the game, Tommy chased him down, but Tigger buttonhooked him, losing himself in the shadows of the perfect hiding place tucked into a darkened notch in the wall down tunnel.

"I gotcha this time, Tommy me boy," he whispered to himself as he hid in the darkness. "Yer all mine."

Neither boy noticed how far they had drifted away from Tigger's post. Neither boy felt the vibration in the tracks, nor heard the coal cars clacking toward the closed trapper door.

It was Tommy who first noticed the puffs of smoke rise along the vibrating tracks, and first heard the distant clanking of cars.

"Tigger! Quick, the cars are coming!" Tommy called out, his words, saturated in fear, choked out breathless.

There was no response.

Panicking, Tommy recognized now how close the cars were to Tigger's door and how far they had strayed from responsibility.

"Tigger!" Tommy screamed in a ragged voice of terror, "you need to open your door!"

Tigger stepped from his hiding place, but seemed frozen in confusion, distracted in the heat of the game. It was only a moment, but it seemed an eternity to Tommy before Tigger turned and ran down the tracks toward his trapper post.

"Tigger, hurry!" Tommy screamed, in sheer panic now.

He turned and ran toward his own station, his heart pounding in his throat, his legs like dead weight, his breath short. Tears streamed down his face. He stumbled, lost his balance, and fell, scraping his hands and face on the rocky floor between the tracks. Bloody, and filthy with coal dust, he pulled himself up and staggered the remaining distance. Sliding to a stop in front of his trapper stool, he looked back over his shoulder and saw Tigger disappearing into darkness.

The iron track rumbled beneath him. The smell of rising coal dust hung thick in the air. The thunder of cars clanking together roared in his ears. His mind raced. It was a big train, four, maybe even five cars, heavy-laden with coal and running out of control.

"Oh no," he breathed. "The hurrier can't stop them."

Tommy placed his trembling hands atop his head and rocked back and forth, staring up the track into the darkened tunnel. His heart lodged securely in his throat, he desperately prayed to see the lights from the coal carts barreling down the track, knowing they would only be visible when Tigger's door had been opened.

He was sweating profusely, his stomach churning. It felt like it would turn inside out.

"Please, God, let me see the lights!" he cried.

But there were no lights. The door remained closed as the out-of-control train of cars raced toward it.

"Open the door!" he pleaded. "Oh, please, God . . . make him open the door!"

Rats scurried past him down both sides of the tunnel, away from the pending disaster.

Nausea flooded Tommy's body. He vomited, then wet himself.

Then came the flash of blinding light, ripping down the tunnel. The clash of steel on steel, hurtling coal, rock, and debris.

Tommy ran. Gripped by intractable fear, he ran harder than he imagined possible. Adrenaline drove his aching heart to a wild pounding beat. Muscles tearing, lungs burning—fighting for air. Tommy ran from the escalating roar swelling into a deafening, exploding crescendo behind him. He ran for safety—but there was nowhere to find it.

The violent compression shockwave from the blast bore down on him, rushing like a locomotive to overtake him in the confined tunnel. In a bullwhip of concussion, it ripped the air with a thunderous *crack* that threw Tommy's small body forward like a rag doll toward the iron rails stretching out before him. Everything seemed to happen in slow motion, even the pounding of his heart as he flew through the air. Instinctively, he swung his arms out in front of him in the darkness to protect himself from the impending collision, catching the iron rails in a bone-cracking jolt an instant before his forehead slammed with a thud into the wooden crossties between the rails.

His unconscious body slid along the ground into a jumble between the two rails. Broken, bruised, and bleeding, he lay motionless, his face in a pool of blood.

A secondary firedamp ignition, fueled by coal dust mixed with the methane-enriched gases, lit up the tunnel in a flash of light, sucking the oxygen out of the air. It left Tommy's clothes blackened, his hair singed, burned his exposed hands and arms that lay atop the hot rails.

Lying unconscious in a heap between the tracks, at risk of the carbon monoxide afterdamp asphyxiation, the minutes ticked by in the pitch-black darkness. He did not move.

Chapter 7

⇟ · ⇞

TOMMY AWOKE TO TOTAL BLACKNESS saturated in silence and suffocating in heat. His head was pounding. The ashen smell of smoke from burnt timber, hot steel, flesh, and coal dust filled the air, making it nearly impossible to breathe. He struggled up to his hands and knees, gusts of searing pain in his wrists ripping up his forearms, throbbing in every part of his body.

Bewildered and scared, Tommy pulled in quick, shallow breaths and sat back on his haunches. Blood streamed down his forehead, his face, and into his mouth, tasting of metallic sweetness. It dripped in a rivulet onto the tracks. He reached up with a trembling hand to touch his aching forehead. Timidly, he felt the oozing wetness of a deep cut, then jerked his hand back when his probing fingers brushed across the hard surface of bone in the open gash.

He tried to keep the tears from coming, but it was no use. "Mamma, help me!" he sobbed in a scratchy voice. "Please help me! I can't breathe!"

There was no answer.

"Mamma, I'm scared," he heaved out in desperation, but his cries were swallowed up in the darkness.

Adrenaline from shock pumped through his broken body, help-ing to numb the pain. Unsteadily, he turned on the tracks to face in the direction of his trapper post.

"I have to get back to my stool," Tommy muttered deliriously. "Papa will be upset with me."

He began to crawl. Enshrouded in darkness, blood, sweat, and tears streaking down his face, he put one hand gently in front of the other, feeling his way along the tracks. The rock and gravel tore at his already bloodied knees, elbows, and hands. Sharp pains shot up his forearms, but he pressed on. Every part of his body ached, but it was not near as painful as the dislocated thoughts of Tigger running through his mind as flashes of the explosion began to drift back to him. The game, the distraction, the awful panic in Tigger's eyes, and the race toward his trapper post.

What have I done? The guilt tore at his very soul. *What will I tell Papa?*

Tommy reached his trapper post. Numb, he groped in the dark-ness for the legs of his stool. Turning it upright, pulling himself up to sit, and trembling, he began to sob, his head between his knees in swelling pain.

"Tigger," Tommy called out hoarsely. "Tigger . . . Tigger!"

Tommy listened for some sign of life, but there was only the sound of trickling water running along the deathly quiet tunnel floor. His hands shook, his body shuddered, his mind ran wild. Tommy dared not leave his trapper stool to venture down the tunnel for fear of what he might find. He leaned over and retched, splattering over his feet what lunch remained in his stomach.

Words tumbled out of his mouth in a blubbering muddle. "Papa . . . I'm sorry . . . I'm at my stool now, Papa!" He slumped back against the wall. "Mamma . . . I need you," he whispered.

For a long time, he sat in pitch-black silence, listening only to the beating of his heart, trying to make sense of the thoughts wandering through his disordered mind.

Then he heard the faint sound of rock shifting against rock.

He sat up, straining to see in the darkness, willing his other senses to come to attention, drawing on every fiber of his mind to concentrate hard, to understand what he heard and felt.

Seconds passed. Had he imagined it?

No, there it was again—the reverberating sound of distant cracking along the tunnel ceiling.

And again, this time louder, the vibrations stronger.

Tommy's heart caught in his throat. The explosion must have loosened the ceiling rock. The tunnel—under three hundred feet of earth, timber, and heavy coal and rock—was breaking up.

He stood to his full height, reached up, and braced his hands up against the tunnel ceiling. He could feel the rumbling, the cracking coming closer.

Maybe he was the only one left. Maybe Papa was dead. Maybe they were all dead. Maybe . . .

"Father in heaven," he pleaded, "I'm afraid! Please don't let me die here alone like Grand. I want to see my family again. I want my mam to hold me, to hug me, to kiss me. I want her to tell me everything will be alright. Please let Papa be alive. Oh, Tigger. Tigger, please don't be dead. I want—"

CRACK!

Tommy jolted his head toward the sound in the darkness. *What was that?* He licked the dried blood on his lips, his heart beating like the wings of a hummingbird as he listened, trying to control his hysteria.

CRACK! CRACK! CRACK!

Tommy's anxious eyes darted around in the darkness. He knocked over the stool.

In the air thick with coal dust, his pounding heart lodged in his throat, he coughed out, "It's breaking apart! Am I gonna be crushed?" Water seemed to be running heavier down the tunnel floor.

There came the screeching sound of rock sliding across rock. And then across steel . . . then . . . was it . . . voices?

"Anyone alive there?" came a faint muffled voice.

"Papa?" he whispered, staring into the blackness. Was his mind playing tricks on him? "Papa?"

"Thomas! Can ya hear me, son? Answer me, son!" A tumble of rocks, creaking timber, and the sound of iron on steel echoed up the tunnel.

Then a warm light burst forth from the darkness, with bobbing shadows coming toward him, flickering off the craggy walls of the tunnel.

"Thomas! Are ya here, son?" Running feet pounded on the rough-hewn tunnel floor, the light bobbing up and down as it came toward him.

Hot tears filled Tommy's eyes. Not bothering to wipe them away, he called back in a scratchy whisper, "Papa!" Peering down the tunnel, he saw Papa's shadowed face, eyes glowing white out of the darkness.

"I'm here, Papa!" Uncontrolled emotion welled up inside him. "I'm here, where I'm supposed to be, Papa! I'm at my trapper post, manning the door, waiting for carts to come!"

Joseph rushed toward his son, eyes wet with relief. "Oh, my boy," he cried. "It's all right!" Joseph swept him up in his powerful arms in a trembling embrace. He rocked Tommy back and forth.

"My son's alive," he called out hoarsely over his shoulder to the men in the tunnel. Tears streamed down Papa's cheeks. "Everythin' will be alright. It'll be all right, son."

Be strong for Papa, Tommy instructed himself. But the tears came anyway. Relief coursed through him. His papa was here. He was not dead after all.

"I wanna go home, Papa. I want my mam." Oh, how he yearned to hug her tight, kiss her cheeks, and never let her go. He wanted to snuggle into the comfort of her arms and let her kiss him as many times as she wanted.

"Oh, Thomas, I was sore afraid for ya!"

"I'm sorry . . . Papa," he choked back the tears, but they were streaming down his cheeks. He tried to tell him more, but all that came between the sobs was, "Darkness . . . afraid . . . alone . . . I'm sorry."

"It's okay, Thomas. It'll be all right, me beautiful little man!" Joseph wiped at his own tears with his sleeve. "Thanks be to God. I thought you might've been kilt, too."

"Killed too?" He looked up, forced now to face his most desperate fear. Hoping he'd misunderstood Papa's meaning. "Killed too?"

"Both da trapper and young Fidget was kilt when the train of cars got away from 'em, ran loose, and crashed." Joseph spoke solemnly. "Fer some reason the trapper didn't open the door. Poor nipper must'a been right behind it when the first heavy car came slidin' down the steep grade and slammed right into it. They was both kilt. It's a terrible thing, son."

Tommy could barely breathe. "No! Papa, no . . . It can't be. Not both of them? . . . Oh, Papa . . . no!"

Tommy buried his head into his papa's shoulder and sobbed uncontrollably.

⊱ · ⊰

After over an hour of waiting, Tommy and his papa were finally allowed to leave the mine. Tommy was exhausted and stiff. It was hard for him to stand. His clothes were torn and tattered. Gray coal dust and dried blood from the gash in his forehead matted his hair and covered his swollen face. His arms burned and broken, his body aching, his scraped and bleeding bare feet moved slowly forward along the railway crossties. Papa's oil lamp cast flickering shadows that fell all around them as they walked.

Tommy felt the death in the air. These acid memories imprinted on his young mind would never leave him. Focusing on his feet shuffling along the track as they approached the bodies of the two boys, he tried hard not to look, but he could feel their presence, their faces seared into his mind. His hands shook, so he clenched them into fists. He could hardly breathe, so he held it in. His heart pounded in his chest. *Don't look.* His mind raced, but as he passed their bodies, he couldn't help himself.

Tigger lay face up, eyes closed. Tommy pulled in a sharp breath. Tigger's ashen face was unmarked and placid. He looked like he was asleep. Though his body, arms, and legs in blood-soaked dungarees were hopelessly tangled in death, he looked peaceful in a macabre sort of way.

Tommy crouched down on his knees next to Tigger's body, realizing he was seeing his dearest friend for the last time.

Lying beside Tigger was Fidget. A steel shaft protruded from his stomach.

They were both past harm now, past worry, seeming utterly fragile in death.

The company had refilled the carts with coal and left these two boys lying there in death, cast aside from the track to make room for work to continue around them.

Tommy couldn't stop the tears from coming. They streamed down his cheeks, dripping onto Tigger's coal-smeared face. He reached down with his sleeve and tried to wipe them off.

"I'm so sorry, Tigger. Please forgive me," he said. "I'm so sorry, Fidget."

Tommy's shoulders slumped from the heavy burden of guilt, shuddering under the weight he would surely carry all the remainder of his life.

For a long moment, he just knelt there, staring at their lifeless bodies. He took Tigger's hand in both of his, and bid him a final farewell, whispering loud enough for Tigger's ears only. "I promise you this, Tigger. I'm gonna do things differently . . . for the both of us." His tears would not stop. Then with a sober nod goodbye, he laid Tigger's hand onto Fidget's, stood, and walked away with his papa.

"What will happen to them, Papa?" he asked as he limped toward the cage that would take him home.

"Their families will come for 'em soon, I suppose." Papa, out of habit, reached for the cage door without looking up and was caught by surprise when it swung into him. He stepped back sharply, eyes wide, face ashen pale.

Out of the cage stepped a purple-faced Candyman Boo Black. His heavily browed solitary eye, lit in an amber fire, glared down at Tommy, while the vein in his forehead seemed to bulge off his balding, sweaty head.

Tommy stepped back toward his papa for protection. But Papa, under the spell of Boo's cold incriminating stare, had stepped away, allowing the maniacal company enforcer to fill the space between them.

Boo turned an intimidating glance toward Papa, who shrank back toward the cage. Reassured by the response, a surly frown spread across Boo's contorted face, exposing his two silver front teeth in a graveyard of rotten ones. He pressed his advantage and spun around on Tommy.

"Playin' games when ya shoulda been payin' attention to your work." Boo spit out his words in a putrid breath of contempt.

His heart jumping for safety into his throat, Tommy glanced nervously about for his papa to intervene. But Papa wasn't coming to his rescue.

"They're all dead because of yer shenanigans, you little fool." With his left hand, Boo grabbed Tommy by his throat and slammed him up against the cage. Boo's right hand held a threatening raised club. "Ya'll get what's coming to ya, boy," Boo snarled.

Fear welled up inside Tommy. He felt abandoned, deserted, scared. But in a moment of transformation, he stiffened his resolve, clenched his shaking hands into fists, lifted his head. He stood up to his full height and stared up into the cold, implacable eye of the cyclops.

There was a moment of hesitation in Boo. Time enough for Tommy's anger to take over. Backed against the cage, Tommy held on, raised his feet, and pushed at Boo's chest with all the adrenaline-driven strength he could muster.

The reaction caught Boo off guard. He momentarily lost his balance, releasing Tommy and stepping backward. In that split second, Tommy was grabbed from behind, pulled into the cage, and the door slammed shut and locked behind him.

Boo smashed his fist into the cage, rocking the heavy metal protective enclosure back. Then, with a cold, barbarous grin, he drug his bleeding knuckles across his face to leave a trail of blood.

The cage slowly lifted up and away from the powerless monster, his vacant, beady eye remaining fixed on Tommy. Tommy shuddered, but kept his eyes locked on the solitary evil eye of the cyclops who stood watching them rise up the shaft until they were lost in the darkness.

Neither Tommy nor Papa said a word in the pregnant silence as they traveled the three hundred feet to the surface. Tears mixed with caked blood and slid down Tommy's cheeks on the long ride upward, with only the sound of clanging cables and water dripping down on them.

Mam hadn't wanted him to even be here. Mam wanted him to use his talents to be something more than a coal miner. Would she have stood by for over an hour when her son desperately needed care for his injuries? Would she have left his friends dead beside the tracks just because the company preferred it that way? He recalled bitterly how Papa wasted his meager pay at the pub when his family was starving. His heart was broken. Mam had stood her ground against Boo Black, placing herself between this vile bully and her children, daring him to knock her down. She had always been there for him. Why hadn't he seen it until now?

Tommy had idolized his papa, all that he did and all that he was. He had wished, above all else, to be just like him someday. But on this day, that changed. The simple clarity of his childhood had vanished. The treasured adoration that had existed between father and son trembled in the balance. For the first time, he saw his papa's flaws glaring back at him. The luster of his childhood hero would never again shine as brightly.

When the cage reached ground level, it was cloudy, and a summer storm poured rain down from the sky. The smoldering sunlight, filtering through the storm clouds, flooded into Tommy's dark, intelligent eyes, wide and pleading. It was the first sunlight he'd seen in a week. Men and machinery were everywhere. Despite the heavy rain, a large crowd had gathered in the yard. There was a collective gasp as Tommy exited the cage, but all Tommy saw was his mam's face—the most beautiful sight in all the world.

"Oh, my darling!" Tears streaming down her cheeks, her arms outstretched wide and welcoming.

He broke free from his papa and hastily stumbled into a broken run, throwing himself into her open arms. In one great sweep, she lifted her boy off his feet. Now both were smeared in blood, coal dust, sweat, and tears as he sobbed love and loss into her shoulder.

"Oh, my dear sweet boy!"

"I hurt, Mamma," he choked out.

She hugged him even more tightly.

He tried to tell her about it, but all that sputtered out between his sobs was, "Tigger . . . Fidget . . . Boo . . . I love you so much . . . I can't do this anymore, Mamma."

That afternoon in the pouring rain of a summer storm, eight-year-old Thomas Wright took the short walk home hand in hand with his mam. That walk marked the end of one journey . . . and the beginning of another.

$\prec \cdot \prec$

Martha, with the help of the homeopathic apothecary, Mrs. Templeton, a mother of six herself, spent the rest of the long day treating Tommy's injuries. Scrubbing off dirt, coal dust, and dried blood. Stitching up the large gash in his forehead, salving his burns, scrapes and cuts with aloe, honey, and other medicinal herbs. One of Tommy's forearms had been fractured, and the other badly sprained. Both were splinted. And after darkness had fallen, Martha put her son to bed. Of course, neither Mercury poultices, proper bleeding off of all the bad blood, nor any of the other more expensive medical treatments were available.

Mrs. Templeton felt sure Tommy's physical injuries would heal in time, of course leaving scars of remembrance for a lifetime. But Martha knew the emotional scars from the tragedy would be different. Those would take away forever what remained of his childhood, changing the way he would look at life. Nothing would ever be the same.

That night, Joseph and Martha sat beside the fire blazing in the hearth and gazed at their four little ones asleep on the floor. Tommy laid curled up in the middle, Georgie and Joey on one side, and Edie on the other. Each of his siblings draped a comforting arm over their older brother to help soften his hurt. His hands and arms were splinted and bandaged to the elbow, feet and legs bandaged to the knee, and his entire head wrapped in a gauze. Only his badly swollen little face and eyes were visible. Martha couldn't help but think of the fragility of life, and the helpless feeling brought on by a tragedy like this, knowing that their lives in this coal miner's village would never be their own.

"They look so angelic, so innocent asleep in front of the fire," Martha observed bitterly. "Life can be so cruel. It can be so brutal. So fragile! It tears at my heart to see Thomas's childhood stolen so."

"It's true enough!" Joseph agreed.

Martha sat quietly, watching her children for a long time in the soft shadows from the flickering flames of the fire that drove the darkness to the corners of the room. She listened in anguish as Joseph shared the details of the death of the two boys, the injury to her own son, the delays of getting needed medical care. Anger rose in her as she listened to him talk of the emotional distress while her severely injured son sat unattended to.

"How could they?" she interrupted. "I waited for hours in the yard with the other mams. Tigger's mam, with no husband to support her, tried desperately to convince the guard to let her take the cage down to her son. She pleaded hysterically for him to show a little compassion, a little mercy to a grieving mam, but the guard was afraid for his job and wouldn't."

Joseph sighed. "'Twas a mess down there. Coal, timber, rock, and debris scattered and piled everywhere. It took more than an hour to dig 'em out."

"But once you were able to break through and get to Thomas . . ." Martha waited, but her husband didn't volunteer the reason for the delay. "We got word that Thomas was alive; Tigger and Fidget had been killed. It still took over an hour before you were able to bring up Thomas." Martha paused. "There were several worried families whose children had not yet come up when we rushed off Thomas to treat his injuries. They were begging to go down into the mine but were all treated badly, including Aunt Mary, who was still waiting for Abe and Zac with no word when I left with Thomas."

"I'm sorry. I knew you'd be upset."

"And you weren't? Tigger and Fidget's mams; Mary and the other families waiting for word of their children were ignored and even assaulted by Boo Black. That monster demanded they 'keep their traps shut,' threatening them with physical harm. He and his thugs corralled us into the corner of the yard and made us stand there in the freezing cold rain with our crying children, babes in arms." Martha looked for a reaction from her husband. "Why?"

"That tunnel is a main artery to that part of the mine."

"I don't understand!"

Joseph took a moment to choose his words carefully. He knew they would not go down well with Martha. "Lettin' the families in would've held up loadin' and takin' the coal out to market. And besides, the pit manager thought it too dangerous for women to be in the mine, so he waited."

Martha inhaled audibly. "Waited . . . for what?" She tried to push back on her building anger long enough to get her question out. "Are you telling me the company just left the bodies of those two dead little boys lying beside the track? Their poor mams, knowing their boys lay dead, waiting in the rain, wind, and cold. Mothers who've . . . already given up their husbands to black lung. One is already dead.

The other's dying on the floor of their hovel, draining the family of every last shilling their boys make working twelve hours a day. They made those desperate mothers wait . . . so they could haul more coal to market?"

"I knew you'd be angry."

"You're not angry?" Martha took another moment to collect herself, her breathing coming fast now. "I don't suppose the company will be paying a single farthing for their burials either."

Joseph took a quick, shallow breath, nervous now, not wanting to tell her the rest of the story but knowing he must. "There's more."

"What do you mean 'there's more'?" She spit the words out.

"While ya had yer hands full tendin' to Thomas's injuries with Mrs. Templeton," Joseph paused, holding his breath, "they found the other missin' bairns."

"What?"

"The company didn't say anythin' to the families waitin' up top . . . 'cause they didn't know where they was. Until they found just what happened to 'em—where they'd gone to."

Martha's eyes were wide with terror now. Her mouth open, but unable to speak as her heart pounded, she held her breath waiting for the other shoe to drop.

"The explosion jammed the headgear making it impossible to operate the cage and bring the miners out. Seems the bairns at the Huskar end of the mine was cut off from the rest. After these little ones put out their lamps to avoid another explosion, they got scared. When they couldn't take the cage up, they decided to take another way out. They crawled up the long steep abandoned Huskar day hole into the bottoms of Nabs Wood." Joseph paused, looking at the last burning embers in the hearth. Martha waited, wide-eyed. "But that thunderstorm we had flooded Nabs Wood."

"Oh, my loving God." Martha's hand came up to cover her mouth.

"The floodin' water poured into the day hole tunnel, washin' all of them together back down the small tunnel. Crushin' them up again' the closed trapper door."

"They drowned?" Martha asked aghast, holding back her tears.

"We lost twenty-six bairns today, boys and girls."

"Oh Joseph . . . no!" Martha threw both hands to her face and held her breath, then whispered, "May God have mercy on their souls."

"They found Abe and Zac, their faces unmarked. Their arms wrapped 'round each other in death."

"Oh no!" Martha wiped at her closed eyes. "Aunt Mary will be left alone with her baby, two little girls, and now Abraham and Isaac gone. No papa or her boys to support those little girls."

"They say the tragedy was an 'awful act of God—an unfortunate accident.'"

"What?" She looked back at her husband in horror. "It was greed for filthy lucre. These horrific, uncaring mine operators and landlords and their unforgivable negligence!"

Joseph tried to put his hand on his wife's shoulder, but she furiously pushed him away.

"Mustn't we accept the way things are and get on with our role in it, Martha?" But Joseph pleadings rung hollow to Martha. "It's been this way for over a hundred years," he went on. "What can we possibly do to change things? Maybe some things can't be explained . . . only forgiven."

"How dare you!" Martha snapped back, horrified. She would have none of it. She was bitter. She was angry. Her heart ached.

Joseph tried to take her in his arms as they sat by the last glowing embers of the fire, but she turned and pushed him away.

"I've spoken to Thomas about his future," she said bitterly. "After this morning he doesn't want to be a coal miner anymore, and I'm determined to do what I must to keep him safe. We both know there are few other choices for work here in this village. But he is no longer digging coal. We must make other arrangements for Thomas and the rest of my children."

"What other arrangements, Martha? There's nothing but the mine. People starve iffen their bairns don't work the mine."

"We'll see about that."

Martha put a protective hand on her swelling belly, her face burning with bitterness. She wanted desperately to give her husband the respect due the patriarch of the family as her religion encouraged, but in her furious anger she had found it impossible to do so.

With hurt filling his eyes, Joseph tried to console her, to help her acknowledge that their family owed something of gratitude to the company that provided their livelihood and a roof over their head. But she could stand no more.

Martha's voice, raised to a fever pitch, was cold and clear in her conviction. "I'm sorry, Joseph, but there must be a special place in hell for these grasping, clawing, arrogant aristocratic industrialists who take unseemly profits whilst unleashing heartache and tragedy. They allow—no, *demand*—these innocent little children to work in their coal mines under horrific conditions as virtual slaves. They take away their most precious inalienable blessings granted by the Almighty Himself. But even that's not enough. After a disaster like this, they prevent grieving mothers from collecting the broken bodies of their little ones, just so these arrogant, impatient, greedy lords and ladies can squeeze every last shilling out of the coal these children died for. I'll see them all in hell before I sacrifice my children on the altar of their vanity, their prideful prejudice, their selfish disdain for those not of their class. Damn them! Damn them all to hell!"

Joseph reached out to put a comforting hand on his wife's shoulder. This time she didn't push him away.

Martha wiped the angry tears from her eyes. "What kind of a mother am I, allowing my innocent little boy's childhood to be stolen so? I was supposed to be helping him find his way. How could I have turned the responsibility to love and cherish my beautiful little boy over to these uncaring aristocrats, without a fight . . . God, please forgive me!"

➤ PART II ❖

Wentworth—Woodhouse, England, 1861

Chapter 8

>• ◄

Cloaked in a hooded cape and pre-dawn shadows, Lydia Kaye walked in brisk strides along the path adjacent to the stream that meandered through the forest. At this early hour, on a late fall morning, the moonlight spread like searching fingers through the sweet-smelling pines, birches, and recumbent sycamores reflecting off complacent eddies in crystalline rainbows of blue-grey mist.

Lydia was a tall, slender young woman with creamy smooth skin, green flashing eyes fringed with ink-black lashes, thick auburn hair, and a gentle curve to her cheeks. She had made it her personal commitment to learn every plant, tree, and animal in the forest. Even after a dozen years in service to Lord Fitzwilliam, she seemed to find something new every day. Unfortunately, after this morning she would no longer be numbered among the more than four hundred and fifty servants attending to the magnificent Woodhouse Mansion on its sprawling estate. She would miss this enchanting forest.

The humility required of Lydia's station as a servant in the great household had belied her natural independent spirit. William Fitzwilliam, the second son of the lord of the manor and a sickly child who had grown into a needy young man, had become increasingly dependent upon Lydia. Still, she never forgot her place as a servant in his mother's household. She never raised her voice in encouragement beyond her station. Still, she and William spoke of many things—books, gardens, friendships. Over the years they had grown as close as any brother and sister.

Then came the day when he wanted something more. William had found it difficult to ignore her emerging beauty, and the beginnings of her attractive swells of womanhood. In the end, it was his unbridled admiration and attraction to her physical charms, colliding with her uncompromising deportment, unbowed in the face of ill-treatment, that would be her undoing.

Lydia felt surprisingly secure when Lady Fitzwilliam had come to her late the previous evening. Indeed, she was even a bit detached from the whole affair, accepting her dismissal with the utmost decorum, but firm in her insistence she had no fault in it.

On the threshold of a new beginning, she had packed up her few personal belongings and taken her walk, as always, through the forest before daybreak. She was determined to put the controversy behind her, to focus only on the path beneath her feet as she picked up her pace. She allowed herself to be captivated by the majesty of the woodland, its sounds and smells and the winding brook running through it. The path didn't judge her, or insist she remember who she was, nor accuse her of attempting to rise above her station in life.

Her heart quickened as she picked up the pace. Puffs of warm breath came rapidly now, in billowy clouds, mingling with the mist lifting off the stream. The path beneath her feet didn't require her to find it interesting, creative, or entertaining in order to share the morning with her. It didn't remind her to be respectful of her betters, despite their unwarranted arrogance and condescension. The path didn't fear her femininity, nor was it concerned that she might cause a social rift, upsetting the balance of the way things were, the way things had always been. And for that, she was grateful.

Only the babble of flowing water and the falling of her footsteps broke the silence as she pressed onward. She was alone in her thoughts, contemplating the new course and direction of her life. She was a bit fearful, to be sure, but with each step along the winding pathway, she had further resigned herself to her change in circumstances and what it would mean for her future.

Without warning, there was a rustling in the underbrush ahead. She stopped silent in her tracks, eyes wide and alert, held her breath, her heart racing.

Thirty feet ahead, Lydia watched a man emerge from the woods onto the pathway. Tall, lean, and as menacing as a whipcord, his flannel shirt was pulled off and tied around his hips leaving him in pants and a tight-fitting long-sleeved undergarment. Each of his strong hands were lifted to hold a pair of the legs of a large buck draped over his shoulders and across his chest in a manner that appeared both casual and deliberate. He was facing down and away, so he hadn't yet realized she was there. As he stepped onto the pathway, the soft rays of early morning highlighted the muscles that rippled across his back and broad shoulders. Lydia's palms broke out in a nervous sweat.

Then he turned, and saw her.

Both stood frozen, staring at each other through the heavy morning mist. Her heart was pounding. She considered turning and running back down the path, but was seized by his commanding presence and penetrating gaze. With unkempt, rich wavy hair, his straight dark brows hung over dark eyes on a sandy complexion with a three-day growth of beard. The lines of his face were so striking, she was reminded of chiseled stone.

He stood silent on the edge of the pathway. Drops of blood had dripped from the animal to flower on his undershirt clinging to him in sweat. The string of his bow stretched taut across his powerfully built chest. And though he had been caught in the act of poaching, Lydia was struck by his calm demeanor and expression of complete control.

He was intimidating in a way she had never experienced in a man before, but somehow, she was not afraid—she was transfixed. Strangely exhilarated. Her breathing slowed. Heat flowed through her entire body as she stood before this man, boldly looking back at her with insolent eyes. For the moment, all time stood still.

In an act of defiance, she pursed her lips, squinted her eyes, and pushed the gingham-lined hood back off her head and over her shoulders to show him she was not afraid. She felt the rush of cool air on her warm face and scalp. She felt alive in the crisp morning air of springtime. She became more aware of the babble of the brook and the call of the nightingale.

He remained utterly still and silent, his attention focused exclusively on her. He seemed determined to continue looking at her without embarrassment or pause. It was unnerving. He was appraising her as though she were the most striking thing he had ever seen. And there was something intriguing beneath those rugged good looks. An inkling of sadness buried deep within, an unexpected, gentle kindness in those intoxicating eyes. His gaze was so distracting she was forced to look down, away from him. Still, Lydia could feel his eyes on her. She wanted to chasten him for his impertinence. His arrogant, brazen behavior, dangerously flaunted despite the penalty of death according to His Lordship's law. But in her rattled state she could not find the words to chastise him. Slowly, she raised her head in defiance, jaw boldly set, and locked her eyes onto his bold, steely gaze.

Then he smiled.

She could forgive him for his uncivil demeanor, but the impertinent smile was more than good breeding could rationalize. It was just too familiar. It sent her heart beating unevenly. Embarrassed by the rush of emotions passing through her veins, she looked away once more. Despite her effort to feel otherwise, she was captivated by this impudent young man.

But when she looked up this time, just as quickly as he had appeared, he was gone.

Chapter 9

$\rightarrow \cdot \prec$

AT FOURTEEN, A GAWKY ANNIE Dale had unremarkable features—dark eyes, heavy toussled brows, cowlicks of black, unruly hair often blown into tangles. But in the morning, she would be heading for her first paying job as a shop's assistant at a butcher's shop nearly a hundred miles away in the thriving town of Barnsley. Her fifteen-year-old cousin, Emma, who had found work at the Barnsley Village Dry Goods Store, would be going with her. Having just finished packing the few belongings she could call her own, Annie was not entirely prepared to leave her childhood behind for a new life. She was still wild at heart, coveting her slingshot, hiking alone through the woods, and above all else, time alone to read.

Annie had been raised in a family of seven girls—five sisters and her cousin Emma all squeezed in together with her mam and Grandma Nanny, like sardines in a can, in their tumbledown house in the quaint hamlet of Handsworth, in South Yorkshire, England. Grandma Nanny, probably the sternest and most stubborn woman God ever blew breath into, had ruled over their household like a hen in a coop without any roosters.

"In the Dale household," Grandma Nanny had declared, "you girls will learn your three *R*s." And they did—pursuing with vigor reading, writing, and arithmetic.

Annie sometimes thought there was but one sensible soul in their whole household, lodged in the body of her tall, skinny cousin, the remarkable Emma Stanley. Emma had been born on the wrong side of the blanket after her mam Hanna found herself in the family way without a husband. And from the beginning, Annie heralded her good fortune to have grown up with this kindred spirit. The two took to their studies well and made a formidable team, each fortifying the other. Emma excelled in the practical side of life, while Annie was more inclined toward cerebral studies of science and literature. She often saw life as a fragment of a poem and her part in it as a character in a story of intrigue.

Grandma's educating had paid off with offers of employment for both Annie and Emma.

Just after breakfast on the first day of her new life, Annie's stern and foreboding Grandma Nanny met Annie in private to share her parting advice. "Use well all that I've taught you," Nanny began. "Work hard. And remember, you are now old enough to have a baby, and God will forever condemn your soul if you meddle with boys."

The precise particulars of the relationship between those two absolutes were never disclosed, leaving Annie in a bit of a quandary. Fortunately, Emma had also received Grandma Nanny's parting advice.

"Oh, it's so delicious to have someone to share secrets with, Annie," Emma confided to her cousin as they loaded their knapsacks of worldly possessions in the back of the wagon for the one-way trip to Barnsley. "What was it Grandma Nanny said? 'We're spring flowers on the cusp of blooming'? What do you suppose that means, exactly?"

"I don't know, really. I can't imagine Grandma Nanny ever having had a romantic experience with a boy." They paused as unpleasant images drifted through their thoughts, causing them to retch one moment and laugh uproariously the next.

"I'm sure with a bit of practical logic, we can figure it out!" Emma adroitly speculated. "Men and women can't be much different from horses, I'm guessing. And we know how horses do it."

"I remember those horses we saw last summer!" Annie frowned, her brows raised over wide eyes. "I wish I didn't. But I especially remember the stallion."

They both giggled.

"I heard from Marylou, who read it in a book," Emma whispered. "How'd she put it? 'A woman can soar to dizzying heights of rapture and wafted oblivion.'"

"Emma! What on earth does that mean?"

"I'm not sure, but she said if it doesn't happen, a woman can have a hysterical breakdown."

Understanding did not come easily, but with unflagging enthusiasm and great zeal for the subject and scientific exploration, Annie and Emma were sure they'd decipher the more salient points. Intent upon uncovering the whole of it, they set aside the subject for further study.

⤝ • ⤞

"We can make over this storeroom closet into yer bedroom," the middle-aged butcher's wife advised gruffly. Annie found Mrs. Jones to be a loquacious employer, blessed with the ability to talk incessantly in an abrasive tone without seeming to take a breath, yet still saying little of importance. Upholstered with the ample curves one might associate with a well-used sofa, she had a cloud of unruly hair so sternly depressed beneath its net that no vagrant tendrils would dare to escape. "This here room will keep ya close to your work. O'course, ya can work off the rent as part of your pay."

"Thank you," said Annie as she walked past the carcasses of cows, shanks of ham, and once-cute little lambs hanging from the low-slung ceiling. "I'm sure it will be just fine."

Of course, Annie had no intention of mentioning her fondness for animals to the butcher or his wife. But she wondered how she would fall asleep with dead cows hanging just outside her door and the musty smell of aging meat wafting through her tiny converted closet. And soon she would learn to butcher those big, brown-eyed animals as they stared up at her.

Despite Mrs. Jones's abrasive welcome, it was clear Mr. Jones, a quiet and thoughtful man, found Annie to be an intelligent girl who worked hard and learned fast. When he discovered she was good with numbers, he shared that fact with Mrs. Jones, who promptly turned over the management of the shop bookkeeping to the capable young girl.

"Of course, ya'll still be expected ta do all the cleanin' and cookin'," reasoned the butcher's wife as she rattled on about her own busy schedule.

While methodically perfecting her skills at the assigned tasks, Annie became an invaluable asset to the butcher shop in more than just her work ethic—she became an expert at studying people. "Annie, you seem to know what customers are looking for even before I do," the butcher commented one afternoon. "You've got a good eye for business."

Even Mrs. Jones had to admit that Annie worked as hard as any two helpers. Within a month she had become an indispensable asset, "my girl-of-all-duties," Mrs. Jones called her. But Annie's constantly increasing workload made for long, tiring days, leaving her little opportunity for socializing, and almost no time for her favorite pastime, reading.

⋟ · ⋞

Many in surrounding company towns often traveled several miles to Barnsley just to frequent the shops because of the varied selection, customer service, and the cripplingly high prices at the company-owned shops in Elsecar, Hoyle Mill, Silkstone, and other surrounding coal mining shantytowns. At Barnsley Village Dry Goods Store, Emma—naturally talkative, inquisitive, and pretty—engaged every customer who came into the store in conversation, asking questions, sharing pleasantries, and flirting with those who deserved to be flirted with. They, of course, returned the cordiality.

"I've seen the way Billy Abbott can't keep his eyes off you at the dry goods store, Emma," Annie shared one Sunday afternoon as they sat in the village square after church. "But then, I suppose I would too, if I were a boy. You're the pretty one. As pretty as a peach, some say."

Emma laughed. "I am thankful we get to see each other so often," Emma said as she leaned back against a tree. "Sharing information about the boys in town."

"Boys flock around you, Emma. I ain't got nothin' to share."

Emma frowned. "I'm sure that's not true."

"Trust me, most boys think me peculiar with my nose stuck in a book." Annie was resigned to the way things were. "I could hardly be called 'pretty as a peach.' I'm more like a fig."

"Oh, Annie! Don't be silly." Emma rolled her eyes. "And the boys don't flock," she said as a boy walking through the square turned, smiled at Emma with interest, completely ignoring Annie's presence, then tripped and almost fell.

Annie raised her brows and snickered, nodding toward the boy to emphasize her point. Emma ignored her, continuing on about Billy Abbott.

"I can tell Billy wants to look at me when he thinks I am not aware of him, and I suppose he wants to talk to me when I am," Emma continued with a bit of wonder. "I know he's smart, but when he's close to me, he can't seem to speak. He's all tongue-tied, stuttering out his words . . . the poor thing. Most every day this week after his shift in the mine, Billy has put on his Sunday best, plastered his hair down, and come into the dry goods store to buy the oddest things—thread, a pencil, shoestrings. Things I know he doesn't need, all just for a minute of conversation with me."

"My point exactly."

After pondering a moment, Emma added, "But ya know, shyness works for Billy Abbott. I'd like to give him a big hug and tell him it's all right. Maybe I'd spend some time with that boy—if he could just get the courage to ask me to!"

"Oh, that poor thing," Annie sympathized. "While you're breakin' hearts, I go about my day virtually unnoticed, spending much of my day in cold storage with dead animals and a carving knife." She paused, pondering the thought, then continued. "Mr. Jones is nice enough. He's old enough to be our gramps, but we have some interesting conversations. His wife, on the other hand, can talk a blue streak about nothing. The truth has little chance of survival around her, and she seems to have a penchant for talking about work—anybody's but her own."

Emma smiled at Annie's dry sense of humor. Never in favor of keeping her thoughts to herself around Annie, Emma often shared them as soon as they entered her head. She was the first to point out the virtues of Thomas Wright as they sat talking of boys in the village square. "Have you seen him? Thomas Wright, I mean?"

Annie nodded. "I have. Mr. Wright comes into the butcher shop from time to time."

Emma grinned. "Well, I think he's handsome. And built like one of those stallions . . . remember? Strong and good-tempered— probably good at other things too," Emma said turning her head with a sly smile.

"Emma, you're shameless!"

"Don't tell me you haven't noticed. You notice everything." She crinkled her nose, deep in thought. "He seems quiet, maybe even a little shy. But there's something about him that almost feels mysterious, dangerous."

Annie rolled her eyes. "Of course, I've noticed him, but he is no boy. He's gotta be at least ten years older than us."

Annie herself was naturally quiet around most people. In fact, she was so quiet she thought herself invisible to most of the butcher's customers. She saw it as an asset. She'd learned that people's character is revealed when they believe nobody is watching. It allowed her to pursue her fascination with human nature without detection. She analyzed those who came into the shop as if they were characters in a book and she was a perceptive novelist. Little went unnoticed by Annie's keen eye and critical intellect. And except for the obligatory exchanges, she was treated like a piece of furniture; people were nice enough around her, but didn't give her much notice.

"I think Thomas Wright is the kind of man it's hard to take your eyes off," Emma concluded dreamily. "If I ever have a chance at a man like Thomas Wright, I certainly won't pass him up for eatin' sweet bread. He seems to have a determination to do something with his life. He's the kind of man you should consider, Annie."

"Don't be ridiculous, Emma. We're not talkin' about a boy here, and a man like that would never be interested in a shy, plain, pimple-faced bluestocking, with no features he'd consider remarkable I might add." Annie turned away from Emma in embarrassment and found herself facing her reflection in a storefront window glass.

"Look!" Annie pointed to her reflection. "Mirrors don't lie. I'm certainly not about to ask, 'Who's the fairest of them all?'"

There was silence between them for a moment, then they both broke into uncontrollable giggling.

Emma frowned. "Don't shortchange yourself, Annie. You understand people and see their character flaws better than they do themselves. And you've got God's greatest gift—a sense of humor." Emma paused.

"You're the sweet roll in a plate full of rye, Annie Dale."

Annie rolled her eyes at this nonsensical sentiment from her best friend. "I ain't no sweet roll, Emma Stanley . . . I'm a realist."

Chapter 10

⤙ · ⤚

I APPRECIATE YOU BEING SUCH A good friend, Edie," Lydia said as they sat at a table in the village square together, sharing sandwiches and tea. "Heaven knows I needed a friend after all those years in service at Woodhouse. But after leaving behind all that controversy, I'm just not ready to meet anybody now."

"What do you mean 'ready'? Edie teased. "You're twenty-five years old. Most women your age are married with half a dozen children." A playful smile spread wide across Edie's sweet face. Punctuated by her prominent widow's peak, her round, dimple-plump smiles always seemed to brighten the room wherever she went.

"You've been back in the village for almost eight months," Edie continued, "and I'm your only friend. That's pathetic. Every eligible man in Elsecar wants to meet you, but it's like you've got this foreboding sign tattooed on your forehead saying, *Don't even think about it.*"

Lydia laughed with a sigh. She couldn't expect Edie to understand. After so much unwanted attention at Woodhouse Mansion, she was in no hurry to be at the center of intrigue in the village. "I'm just not interested right now."

Edie, finishing off the last crumbs of her sandwich, replied, "Well, if you don't watch out, you'll end up an old maid."

"That suits me just fine. I'm not planning on marrying."

"Just do me a favor and at least meet my older brother."

Lydia rolled her eyes.

"After all, you have so much in common. You're both over-educated. Both think too much. Both are highly adept at driving off marriageable possibilities. Both recluses at risk of living alone the rest of your lives, and you both consider a surly disposition a prized virtue."

"Really, Edie, that seems a bit harsh."

Edie shot her a knowing smile. "You don't even have to talk to him if you don't want." Then offered with a giggle, "Most of my girl-friends think he's gorgeous to look at."

Lydia laughed. "Perhaps I'll consider it." Eager to divert the con-versation away from boys and courting, Lydia asked, "How did a coal miner's son end up Engineering Manager for the Silkstone & Elsecar Coalowners Company?"

⤙ • ⤚

Edie noted Lydia's attractive brows raised in interest and cleverly cast her line in that direction. "Our mam."

"I'm guessing she wasn't happy about her bright little boy going down into the pit, but what's her role in it?"

Edie chuckled. "Mam is a force to be reckoned with," Edie nod-ded. "Papa won that first battle—Papa, Grand, and way on back were all miners. Come to think of it, that may have been the last win for Papa." Edie began to reel Lydia in with an innocent smile, giving her a quick summary of Tom's life story: going down into the mine as a child, the death of his childhood friends. "We lost twenty-six children from the shantytown on that day."

"Oh, my sweet Jesus." Lydia put her hand over her mouth to stop the emotion escaping.

"They were children Tommy knew and worked with every day in the lower regions of the mine." Edie paused, remembering back. "Eleven of them were girls. It was horrific for the whole village."

Lydia's brows pulled together and she frowned in disbelief at such a tragedy. "I suppose that settled the argument."

"Oh, it did for a long while. Tommy stayed home and kept to himself. For weeks after he didn't speak to anyone. Most every night he woke up with nightmares."

Edie pulled back the memory of those difficult days. Tommy was despondent, angry, and confused and felt inconsolable guilt after the tragedy. It was several weeks before he even began to read again. But when he did, he read anything Mam could find, in part to divert his mind from death and tragedy. He was fascinated with scientific discovery, learning the why, the way, and the wherefore of things. The weeks rolled into months, then into a year, then another year, and into adolescence. "He didn't associate with anyone outside the family. Frankly, most all of 'em around his age were dead. He was a voracious reader—still is. Mam found books for him to read in the most obscure places. Eventually, as the years rolled by, traveling all the way to Sheffield for books on science, mathematics, and engineering. Tommy devoured texts whose titles I couldn't even pronounce, memorizing them from cover to cover."

"I can imagine how those tragedies must have hardened your mam's resolve."

"After the disaster, Mam organized several of the mothers of lost children and launched a campaign to push back on children in the mine. The backlash caught on and Lord Ashley, a local aristocrat, pursued legislation to restrict child labor all across England. It took almost five years, but with a bit of help from Queen Victoria, Lord

Ashley, a member of Parliament by then, pushed a law through Parliament keeping girls out of the mine and boys younger than ten." Edie paused, thinking back. It had lit a passion in Martha. She and a few angry mams, who were not afraid to rock the boat, found it was possible to make a difference. Mam was inspired. Maybe there was a chance to change the cycle that kept children uneducated and forced them into industrial slavery. "It was a sore spot for Lord Fitzwilliam, and I'm not sure he forgave Martha for it. But Tommy caught Mam's fire." Edie remembered how it had given him a real focus for his studies. How he had become obsessed, distilling Mam's passion to a crystal purity.

"What about your other brothers? Was your mam able to keep them out of the mine?"

"She did for a while. Even when Papa got sick, and began missing work, exhausted with coughing attacks. It turned out to be black lung. More and more, the money he made was not enough to put food on the table for our family."

"I'm so sorry, Edie. Did your brother have to return to the mine?"

Edie nodded. "We were living on company credit and close to starving. Tommy was forced back into the mine working side by side with Papa. Hunger is a pushing thing. Still, thanks to Tommy, Mam was able to hold both Joey and Georgie back. But with Papa increasingly out of work, Tommy became the primary breadwinner in our family."

"That must have been brutal for a young boy."

"It was an awful time for him. All of us, really. But without Tommy . . ." She trailed off. "Honestly, we would have starved to death. But even so, Mam made sure he kept up with his studies."

Edie frowned. "My brother is brilliant, but for a long time, all he did was work to provide for the family and study. He's still estranged from everyone outside of family, even when he is at the mine."

Lydia sat back pondering. "Your poor brother. I know a little about being alone with no one to lean on."

"He still blames himself for his friends' deaths."

"Why?"

"It's complicated, but he's carried that cross all these years. From the beginning the driving force to educate himself, fight against child labor, and his push for safety measures has been a kind of redemptive penance. His struggle to overcome his shortcomings—it's his kind of reparation for his transgressions . . . what he sees as his sins, I suppose."

Lydia was silent for a long while. "What a heavy burden for a child to carry." She looked at Edie, "There's no talking him out of it?"

"Believe me, we've tried. Tom is obsessed with changing the way of things. Willing to put himself at risk to do something about it. He stays just beyond the reach of the company's ire."

"I know something of the miserly Lord Fitzwilliam. These are arrogant men. Dangerous men."

"Tom's been too busy achieving his ends—too busy doing it to spend much time listening to those who say this is not how it's done."

Lydia's brows knit in concern. "That's inspiring, really, but how could the son of a coal miner convince a company of powerful men to spend money they don't want to spend?"

Edie grinned mischievously. "Trickery! At least that's what Tom calls it."

"What's 'trickery' supposed to mean?"

"Well, there was the time Tommy had a run-in with Candyman Boo Black. He—" Edie was suddenly distracted.

Lydia followed her gaze to a pair of figures in the distance coming across the square.

Edie smiled as though their clandestine afternoon tea had just been discovered. "Well, well, will you look at this? My little brother Joey and my husband Andrew." Edie laughed as Andrew, her tall, thin, sallow-faced husband, leaned down to give her a kiss. "It's a family convention."

"Mam asked me to pick up a few things after work," Joey chimed in. Then, seeing Lydia, asked, "So, who's this, Sis?"

"You've not met Lydia Kaye, have you? Neither of you." Edie introduced both to Lydia. "I suppose you better pull up a chair and join us."

"Hello, Lydia. I'm Joey—Edie's best-looking brother."

Lydia smiled at his boyish charm. "Nice to meet you, Joey. And you, Mr. Hobson."

"Stop it, Joey." Edie waved him away. "This is not about you."

Lydia's irresistible smile lit up her face. "Your sister was just telling me the story about the day your brother ran into trouble with Candy-man Boo Black."

"Which day would that be?"

Lydia laughed. "There's more than one?"

"I was *going* to tell her the story," Edie interjected, "about when you and Tommy had that run-in with Boo Black at the canteen."

"Oh, the infamous boy-genius story. Yea, I was there. I can tell it. In the end you'll see why a beautiful young lady like yourself should not bother with a recluse like Tommy." Joey looked at his sister, who, while rolling her eyes, nodded her assent for him to continue with the story. "It's best you stick with a sure thing like me."

Lydia smiled, amused by his flirtations.

Joey leaned back in his chair and launched into his dramatic telling.

⊰ • ⊱

"Tommy had just turned fourteen," he began. "I was nine. It was a dark time for our family. Papa had been sick, his breathing ravaged by black lung. So, that left Tommy to take most of the family's financial burden on his shoulders, spending long days digging coal while the rest of us scavenged and picked up odd jobs around the village. Papa spent more and more of his time at the pub. But Mam, who by then held the purse strings, wouldn't finance his drunkenness," Joey smiled, "and she wouldn't dare let him get drunk on credit."

Lydia's brows squeezed together, then, as she raised them, her emerald eyes filled with compassion. She seemed vaguely wistful as she listened. "Your mam must be very strong to keep your family together under those trying times."

Joey only nodded and continued with the story that had become folklore in the family. "So, one afternoon, Tommy and I were sitting at the Silkstone Canteen lunch table, where I often brought him lunch from Mam. Tommy had a book on his lap."

"*The Thermodynamics of Ventilation* by Parkin Jeffcock," Edie interjected. "Mam still has it on the fireplace shelf at home as a kind of a monument to Tommy's new start in life. She found it in a Sheffield rubbish sale, of all places. It's water-stained, the pages are worn and tattered, but Tommy says it's filled to the brim with what he calls 'golden nuggets.' All I see is a lot of gibberish and really complicated math. I can't make heads nor tails of it."

Joey rolled his eyes. "Thank you for that bit of enlightenment, Edie."

"That's a long way from reading children's books by candlelight." Lydia smiled.

Joey cleared his throat to continue. "He was going over his drawing, mumbling through an improved ventilation system in the mine, while I was munching on my bread-and-cheese lunch. Firedamp, afterdamp, flammable gases—that kinda thing. Poor ventilation can make conditions perfect for a spark from a hammer, candle, or the crash of a coal cart to cause an explosion. Trapper doors, all operated by little children, were supposed to keep the air flowing so they didn't have this dangerous mix of gases. Tommy would always tell me it was a wretched solution to a very dangerous problem."

"Really, it seems unfathomable," Lydia interrupted. "There must be better ways."

"Of course there are, and Tom was discovering that. He found the problem could be solved by relocating the ventilating furnace, or better yet, replacing it with a more efficient Guibal fan which he had just read was under study in France. With a better design, the number of trapper doors could be cut dramatically, or even eliminated. He also found the company would save money, because they wouldn't need unreliable children and they'd avoid expensive disasters that caused workers to lose their lives."

"My brother really is a brilliant engineer," Edie interjected. "Some men of science grapple with problems right up to the edge of darkness, searching for answers to questions most of us don't even see. But Tommy is a genius. He jumps right into the middle of the blackest unknown uncovering secrets that God only knows, then somehow he finds his way out of the darkness."

Joey recalled his older brother back then, the boy genius lost deep in thought, imagining chimney furnace fans, their vanes turning in the direction of the hazardous exhaust gases to clear them out. Painting a picture of the air swarming with invisible poisonous gases passing through galleries, over headways, down chutes, under arch supports, preventing crushing pressures of the heavy overburden of rock, coal, and earth from collapsing the tunnel . . .

"Tommy was always getting lost in his thoughts—what Mam calls his 'fertile imagination.' He'd just stop talking midsentence sometimes. That's what happened that day. I was saying his name over and over, trying to get his attention, 'cause Candyman Boo Black had just walked in and he was coming over to our table. Tom looked up from his sketches. I'm sure my face looked panicked. Candyman Boo was standing over us, angry as always, looking down on Tommy." Joey paused.

"I know the monster," Lydia said in disgust. "I've seen him with Lord Fitzwilliam. A big, terrifying, sadistic brute. Rumor around Woodhouse was that he got that scar on his forehead and lost his eye in one of his father's drunken fits and when Boo was old enough, he killed him."

"That's him," Joey bitterly confirmed. "Most every miner has a story of somebody being brutally clubbed, sometimes to death, by Candyman Boo Black."

"He's utterly sick." Lydia shuddered. She smoothed out her skirt, and looked back at Joey, curiosity betraying itself as she asked, "So, what happened with your brother, Joey?"

Joey continued, "Boo looked down on Tommy, wrenching up his ugly face. 'Hey, boy, don't ignore me,' Boo said. He had this yellow spittle hanging in the corner of his mouth. 'Next time I be gettin' your attention with a club upside the back of your head.

"Tommy looked around the room confused, wondering just what his transgression had been.

"'What's this here book, boy?' asked Boo. He yanked the engineering book off Tommy's lap. It was so heavy, Boo nearly dropped it. He stared at the open page, clearly baffled by the sines, cosines, and Pythagorean theorem equations. A ladder of creases sharpened up his forehead, running halfway up to his crown.

"I could see Tommy's cheeks flush as he watched this ignorant lunatic trying to decipher the meaning of the words. 'What are ya readin' this here fer?' asked Boo.

"'I'm learning a bit about things that might help production in the mine,' said Tommy. He was nervous. As soon as he said it, it was obvious he wished he hadn't.

"'That's not your concern, boy.'

"The men in the canteen had begun to gather around us, gawking. 'He's quicker 'an any us miners 'bout this kind of stuff—science and all,' said this one talkative young miner. 'Da safety of the mine,' said he. 'Maybe ya should listen to—'

"But before the boy could finish, Boo took one step toward the miner and with this wide sweep of his balled-up fist, he backhanded him across his face. His jaw twisted, and blood spurted from his nose and mouth. He fell back hard, slammed his head against a table, then fell to the floor."

"Boo Black stepped up to the boy lying there, unconscious, and started kicking him hard, driving his heavy leather boot into the boy's cheekbones. I still remember the sickly sound of his bones collapsing. 'That's none of your business, boy,' he was saying. And then he turned to one of the other miners, 'You—get 'im outta here!'"

Joey swallowed hard, took a deep breath, blew it out, and continued. "I was sick to my stomach, and my heart was pounding in my throat. But I was frozen to my seat." Joey winced as he remembered the sound of the boy's inert body being dragged across the blood-smeared floor, breaking the silence in the room. Boo's fiercely black eyes, blazing with evil, flashed into his mind, as it did each of the pale-faced, wide-eyed miners.

"I vividly remember what Boo said next. 'Ever' mornin' when that boy looks in da mirror, he'll remember da day he crossed me. He'll never forget me, or da day I became da most important man in his life.'" Joey spat out Boo's words. "One by one, the miners took a step back and stared at the blood smeared on the floor. My heart was sick with fear, knowing every time I saw that boy, I'd remember too."

Shock registered on Lydia's face as she glanced at Edie. Andrew's arm was draped protectively around her. Lydia turned blankly back to Joey, silently processing his story.

"Then Boo turned on Tommy," Joey said, his hands clenched into fists. "I was sure Tommy was a goner, but right then, the pit manager stepped into the canteen.

"'What's goin' on here?' he asked. Boo scowled at the pit manager, who took the book from Boo and thumbed through the pages. Then he glanced down at Tom's sketches on the table. 'This kind of stuff's too expensive. Best put it aside and not concern yerself with it,' said the manager. And he put the book back down on the table in front of Tom, who was sitting silent and red-faced. Then, just like that, the manager walked out."

Lydia breathed a long sigh of relief. "He didn't say anything to Boo Black?"

"Nothing," Joey said. "Boo stepped up to Tommy real close like, and whispered, 'Ya don't need to know nothin' but diggin' fer coal, boy.' I could smell the sweat, putrid breath, and alcohol. Ya can leave the thinkin' to yer betters.' Then Boo sneered, 'I hear yer old man ain't doin' so well, black lung and all. Word is, he ain't long for this world. Maybe best to put 'im outta his misery?'"

"Oh my, he was taunting your poor brother. What happened?"

"All I could think was, *Tommy, please don't say anything*, but it was too much for him. He grabbed his book, stood up from the table, fuming mad, and headed for the door. Tommy was tall, but all knees and elbows back then, and he clumsily bumped Boo on the way out. Boo, with clenched fist, had been preparing to whack Tom too, but he was knocked off balance into the table. Tommy, in long brisk steps, was already at the door by the time Boo recovered, and was shouting out curses. And he was already gone when Boo called out, 'Big mistake, boy. Ya'll be payin' for this.'"

Joey could still see the faces of the other miners in his mind's eye, none uttering a word, their nervous eyes darting between the Candyman, the door, then down at the blood on the floor.

"Poor boy must have been beside himself in frustration for losing his temper."

"He was absolutely certain his life was over."

"So, what happened to Tommy?" Lydia demanded, anxious now.

"That night at dinner," Edie picked up the story, "we could tell Tommy was upset. He sat brooding, pushing his food around on his plate without eating any of it. He avoided eye contact with Mam.

"'What happened today?' Mam calmly asked.

"For a long moment Tommy just sat there, considering the question while all of us looked on. Then he whispered, 'I threatened the Candyman. Then I walked out on him while he cursed me.'

"We all stopped eating and stared at Tommy, then at Mam, then back to Tommy."

Edie's eyes widened. "'What were you thinking of?' said I foolishly. 'Papa's gonna kill you when he gets home from the pub.'

"'He ridiculed me,' said Tommy as he looked at Mam. 'He threatened me and Papa.' Tommy looked up at Mam, his brows pushed together, his quivering lower lip of his frown pushed out. 'So, I lost it, Mam. I'm sorry. I know we desperately need the money . . . but they don't own me. And if I have anything to do about it, they never will.'"

Lydia was on the edge of her seat now. "What did your mam do?"

"Mam didn't say a word. Everyone at the table had seen families starving in the streets. You know, the ones with bloated bellies, the yellow pallor, drooping shoulders, and hopeless eyes. And when they die, the company coroner never calls it for what it is, death by starvation. It's always consumption, heart failure, or some other disease." Edie paused, remembering. "If it was near impossible to feed our family on Tom's pay, it was absolutely impossible without it."

Lydia's nostrils flared. "Life in this village can be so brutal. Sometimes, I think these company men should be shot."

Edie continued soberly. "Suddenly, there was a knock at the door. A boy had come with a message for Tom to be at the pit manager's office first thing in the morning."

"Oh my!" Lydia's brows quickly raised into her forehead as she brought her hand to her mouth.

"All at the table stared at each other." Edie wiped at her dry lips at the memory. "Tommy looked like he had the weight of an entire cart of coal dumped on his back. He didn't get much sleep that night. None of us did."

"What an incredible story," Lydia slowly shook her head. "What happened to Tom the next day?"

"Look, Lydia," Edie said, in a tone that brooked no disagreement, "Andrew and I are going to have a summer picnic this Sunday after church, and I want you and Joey to come. Then Tom can tell you the rest of the story."

Chapter 11

≻ • ≺

G OOD MORNING, I'M HERE TO see Miss Thompson."
Tom handed over his papers to the Woodhouse guard, who
sat high atop a spacious gatehouse's second level overlooking the enormous double gates.

Tom stared at the magnificent gatehouse, the awe-inspiring mansion in the distance, and the immaculate grounds surrounding it. The
stunning scene left him wondering how one man in good conscience
could justify such lavish personal expenses when the safety systems in
his mines were so lacking.

The guard, having completed his review of Tom's papers, pulled up
his spectacles and looked down his nose at this shabbily dressed visitor.

"Take this man to the kitchen," the guard instructed. "Run along
now, Tobin. Check with Miss Thompson afore ya let loose of 'im."

Tom was escorted by Tobin along the path to the kitchen delivery
door. "Everyone in town knows it's a grand place," Tom muttered,
"but to stand, awestruck before this immense mansion, is truly something to behold."

"That it is," Tobin responded.

Tom looked out at the magnificent grounds, exotic plants, immense gardens, lawns as far as the eye could see—everything meticulously cared for. And beyond it, lush green forest along the edge of the lake. It was a forest he supposed he knew better than most.

The mansion was so large, it must've taken a whole day just to walk from the first room to the last—119 bedrooms, fifty-seven bathrooms, more than three hundred rooms in all, needing cleaning and polishing every day. There was an acre of stables, fifty-eight horses, and over 450 servants. Probably the most impressive estate in all of Europe—for Lord Fitzwilliam, Her Ladyship, their three children, and the woman people in town called "a battle-ax of a mother-in-law."

At the kitchen's service entry, Tobin told Tom to wait at the giant door while he went inside. He returned with a gregarious, heavyset woman.

"Thank you, Tobin," said the woman with a frown. With a commanding flick of her wrist, she dismissed the boy, who hurried off, clearly relieved to avoid further scrutiny. While drying her hands on her apron, she tilted her head sideways, one brow raised to size up her visitor. Then with a jolly laugh, she reached out and took Tom's hands in both hers and exclaimed, "I'm Tilly Thompson." Her wide smiling eyes lingering on him a moment too long, she added, "Oh, my! Aren't you a beautiful thing. Martha done good!" With a wide sweep of her arm, she invited him inside.

"You know my mam, Miss Thompson?" Tom smiled curiously as he stepped into the enormous kitchen.

"Goodness, yes! I knowed Martha since we was wee ones. For years, your mam and me slept in the same bed down in the servant's dungeon."

"Really?"

While Tilly regaled some nostalgic memories, she ushered Tom farther into the vast hold of the open, high-walled kitchen with vaulted ceilings. More giant oak doors led to an enormous pantry and its adjacent scullery utility room.

Tom's surprised eyes under raised brows were flooded with the sheer excess of incredible appliances, gadgets, and other offerings in this grand kitchen. It seemed to have two of every imaginable appliance known to man, polished to shine, with lamps in every nook and cranny and beautiful white linens everywhere he looked. He heard the clanking of silverware and clatter of plates being washed by the scullery maids. And the food—he had never seen or smelled such food like this before. It was beautiful just to look at. Plates and platters were filled with pastries, cuts of prime beef, and ceramic bowls filled with gravy and piled high with mashed potatoes. Wooden bowls of salads and freshly baked bread were scattered throughout the kitchen, and Tom counted half a dozen copper tins of soup. This was more kinds of food in one place than he had seen in a lifetime.

Tom rubbed his hand over the smooth finish of the walnut preparation table. It was as long as a colliery drift. He felt the involuntary pangs of hunger rising inside him.

Tilly, seeing the question in Tom's eyes, offered, "Amazing, ain't it? And we can thank you coal miners for it all."

"It's certainly magnificent!" Tom couldn't help but think of his poor papa lying in bed all day, wasting away with black lung, and not a shilling from the company for lost wages. He could feel the heat rising to his face. "But don't thank me. I haven't dug coal for years, thanks to Mam.

He paused for a moment to look at the food before him. "What must it be like to have food prepared for you like this every day?"

he murmured in awe. "Looks like you're preparing for a feast, Miss Thompson."

"Thomas, please call me Tilly. Everybody does," Tilly laughed. "And this is not for any feast."

"Oh." Thomas was in a bit of a daze at the sheer volume of food.

"Is that your growling stomach I hear?" Tilly asked, smiling, and picked up a plate. "I know it must look overwhelming to ya," she said soberly. "It's been hard for Martha and your family, hasn't it? Here, let me give ya a piece of peach cobbler. I'll bet you've never tasted it before."

"Oh no, Tilly. I don't want to take ya from your work."

Tilly laughed. "Hardly. These are the leftovers from Her Ladyship's luncheon. Everyone has already gone. Most of this will be thrown out, slop for the pigs."

Tom looked at the food dishes spread out on the long table. There was enough food left over from the luncheon to feed his entire neighborhood for a month. And in a style the desperate mothers of dozens of emaciated children could not even have imagined. The inequity hit him like a kick of a mule in his stomach. He knew disparity between the rich and the poor existed, but it was a shock to actually see it so starkly. *This is a crime*, he thought, frowning. Mam had spent her days worrying about how to put the next meal on the table when he was a boy, but she had worked here once. What was it like for her, to have this vision lodged in her mind while her children were growing up malnourished and even starving at times?

Tilly turned around, wielding a large slice of peach cobbler on a plate. She added a dollop of fresh whipped cream and fruit from a terrine, then handed Tom the plate along with a freshly baked roll of bread, still warm from the oven. The perfume was so sweet, his stomach growled in want.

"Please, sit." She pushed him into a chair at the table, set with silverware and a soft napkin, and served him on a porcelain plate. "Eat up."

"Oh my, Miss Thompson, thank you!"

Tom didn't know what to do. He looked around to see if anyone was watching. Then he picked up the fork and took his first bite. The crust melted in his mouth, and the flavor of the sugar-sweetened peaches burst across his taste buds in a delight unlike anything he had ever tasted in his life, the first bite sliding down his throat like rapture.

"This must be what food tastes like in heaven," Tom sighed.

Tilly grinned. "Better, I'd wager." She laughed raucously at her own humor.

After Tom had eaten for a few minutes, Tilly set in with questions.

"So, you're the mechanical wizard everyone's been talkin' 'bout. They say ya can fix just 'bout anythin'. Rebuilt the entire ventilation system at Elsecar, I hear, and just 'bout everything else to keep those poor buggers safe down in those awful pits." Tilly paused, then whispered with her hand to her mouth. "Not sure how ya got that old skinflint to cough up the money."

"Don't believe everything you hear, Tilly."

"So, how's my girl Martha? I ain't seen her in far too long."

"Mam's doing well. Rules the family with a velvet-gloved hammer."

"Hmm. Not surprised," Tilly said. "I remember the first day little Martha walked through them great gates like it was yesterday." She stared off in the distance, nostalgically. "I was a scullery maid and watched her from the kitchen window when she come. Barely six, she was—prettiest little thing ya ever seen. I'd just turned ten meself." Tilly paused in recollection. "She walked up that very path, across that there lawn on her wobbly little legs, her face gettin' redder with each step toward me. She was doin' her best to stop them tears from streamin' down her little rosy cheeks." Tilly paused again.

"Every night for more 'n a month, that little girl hid her face in her pillow and sobbed herself to sleep," she said, then began to look out the window vacantly. "Her mam passed while birthing. Then her baby sister died, too. So her drunken pa, ole Frank, sold Martha into service here. She had begged him not to, asked to let her take over care of her brothers. 'Mam taught me good,' she'd say. But ole Frank wouldn't budge. He sold her off and put dem little boys right down in the coal mine. He promised to bring 'em by, but he never did. Near broke her heart, poor thing. I suppose she been a good mam to ya, determined to make up for her losses."

Tom's heart sank as he heard about Mam's sorrows. "I didn't know." He ached for her, knowing she had held all that inside her for so long. It explained a lot about her passion to give her children a better life. He wished she were here right now so he could hug her.

"She doesn't talk about her time here," he said, frowning, his lips in a tight line.

"No, I suppose Martha wouldn't," Tilly whispered, a look of sadness in her eyes. "She quietly set aside the loss of her childhood, but it took a long time to divorce herself from her family."

Tom reflected for a moment. "How did my mam learn to read?"

"It was young Miss Lily Fitzwilliam who taught Martha to read. Hooked on readin' from the start, she was. Often coaxed Miss Lily into sneaking books from her papa's great library so Martha could read by candlelight at night." Tilly chuckled. "She was a pretty young thing, with the determination of a firebrand! A tall, slender girl with creamy, smooth skin, blossoming with attractive feminine curves. But late one night, Martha was caught in the library by Lord Fitzwilliam."

"What happened?"

"She wouldn't say, but she was real upset when she came back to our room, her face all red and puffy. That very night she packed up all 'er belongings, put 'em in that small ship's locker she'd brung ten years before, and left afore daybreak." Tilly looked on sadly. "Not long after, we heard she'd married yer papa, and then we heard you were born. But we never saw her here again."

The great hall door that separated the stairs up to the main dining hall slammed. A woman entered the kitchen in a swirl of commotion. Clearly the lady of the house venturing into unaccustomed territory, Lady Fitzwilliam was still dressed for the luncheon in a colorful Paris original improvement skirt supported with crinoline petticoats, and adorned with richly decorated pleats. Her French hip-length jacket, open in the front to reveal petticoat laces supported by whale bone stays, was complete with split sleeves of flowing silk that swayed in the breeze as she flounced toward them in high curved shoes with buckles.

Lady Fitzwilliam's hair was piled high with a ridiculous headdress, the weight of her hair forcing an exaggerated posture. She was flanked by two lady's maids trailing slightly behind. Both were dressed in stiffly starched white linen, and they tended to their mistress's every need as they cleared the way for her to walk through the risky obstacle course of the kitchen.

"Tilly, what do we have here?" Lady Fitzwilliam pointed an elaborately gloved accusing finger toward Tom.

"This here is Thomas, Martha's oldest," Tilly said, turning toward Tom. "He stopped by after work at the mine to look at rebuildin' our ventilation system."

This revelation caught Her Ladyship off guard, bringing a look of disturbed surprise to her face.

"How fascinating!" She was clearly being sarcastic. Seemingly at a loss for words, she looked down her nose at him in affected disinterest, appraising him as she would an animal in a corral, then she turned back to Tilly. "Make sure he washes up and gets the dirt from under his fingernails before you let him into the dining room. I don't want him touching anything. And don't let him linger around any longer than he must." Without looking back at Tom, she added, "And do something with those clothes. Really, Tilly!"

Tom blinked, the slightest glint of mischievous amusement in his eyes as if he were witnessing the arrogance of a spoiled child.

"Why, madam!" he exclaimed in mocking innocence. "These are my Sunday best!" Tom did not smile. "Is there not an advantage for an Englishman to remain uncorrupted by the sophistication of . . . French fashion?"

"Humph!" Lady Fitzwilliam swirled around, clearly irritated at Tom's mocking command of the English language and backhanded insult. She walked back the way she had come. Her two lady's maids struggled to keep up as their matron strode out muttering, "How dare that disrespectful lout!"

When Lady Fitzwilliam's entourage had gone, Tilly snickered.

"Some folks call her the Dragon Lady!"

Tom grinned. "I'm guessing Her Ladyship would prefer to be caught in the state of nature rather than be accused of the crime of being out of fashion."

"'The crime of being out of fashion,'" Tilly repeated under her breath. "Very good, Thomas. Follow me. I'll shows ya Her Ladyship's concern with the ventilation system. It gets awfully hot down here in the kitchen." Then Tilly whispered like she was at a funeral service, "I don't suppose she would be concerned, but for the bit of heat seeping up into the master dining room—gets mighty toasty up there in the summer months."

Tom nodded. "Let's see what we can do to draw off some of that heat upstairs, and give you a bit more cool, fresh air down here as well, shall we? Maybe we can warm it up a bit in the winter too? Whatta ya say, Tilly? Think we can we make Her Ladyship happy?"

Tilly chuckled. "Ain't sure that's possible."

Chapter 12

＋・＋

Despite the distance, many from the surrounding company villages often frequented the Barnsley Butcher Shop, where prices were a bargain, selection good, and service next to none—especially compared to the Wentworth Company Store, not so secretly known as the Worthless Company Store. Annie observed that Tom Wright was a frequent visitor. He was quiet, much older than she, and always polite and considerate. His bearing alone commanded respect. In his own self-assured way, he seemed to have an air about him that suggested he knew something that others around him didn't. Annie noticed his subtle, self-deprecating sense of humor—an endearing trait that made both men and women smile respectfully. Of course, it didn't hurt that he was attractive, an attribute which didn't take a particular reasoned analysis from an observant young girl to deduce.

One afternoon, Mrs. Jones directed Annie to load Mr. Wright's wagon with his purchases for the long trip back to his family's home in Elsecar. As usual, he was reluctant to let her carry the heavy load.

"Please, Miss Dale, let me take that block of ice." He reached out and picked up the ice block before she could object. It was a simple act of kindness, but one seldom observed by the invisible Annie

Dale, who was constantly asked to lug heavy items for customers, then retreated into cold storage to butcher slabs of beef and shoulders of pork.

"Thank you, but I . . ."

"That's really not necessary, Mr. Wright," Mrs. Jones interrupted. "Annie can get it for you. That's what she's paid for."

Annie noticed the faint barrier in the butcher's wife's words and immediately went quiet.

"Just look at her, Mr. Wright," Mrs. Jones whispered in her foghorn voice just loud enough for Annie and the other customers around her to hear. "That girl is a regular beast of burden."

Tom could see the young girl's face burn with embarrassment, but she was reluctant to demand the respect she clearly deserved. Annie looked up as Tom clenched his teeth behind tight lips of irritation. He turned to face the butcher's wife.

"Really, Mrs. Jones?" Thomas replied with a steely edge to his voice.

"Don't get me wrong, Mr. Wright. Both me and the mister agree she has become our girl-of-all-duties. Mighty helpful, for certain," Mrs. Jones shared. "But she spends all her free time readin' who knows what," she whispered, looking around to make sure all were listening to her witty remarks. "And what self-respectin' man is ever gonna be interested in a big-boned, plain girl like her, with her nose always stuck in a book? A regular Amazon woman," she chuckled.

Annie's dark, intelligent eyes shined below her thick, dark brows, scrunched close together as she took in the insult with tight lips. The pimply-faced adolescent fiddled with the tattered ribbon that pulled her hair severely back into a ponytail cascading over her heavily muscled shoulders from the hard work of heavy lifting.

"Sounds like you're very fortunate, Mrs. Jones!" Tom snapped.

Annie's surprised eyes flickered up at his irritated response.

"What must it be like to have an asset like Miss Dale around?" he continued. "I'll bet she could be the queen bee around here, run this entire place without either of you."

Tom paused, his cold, simmering anger rising. By now, all was quiet in the shop. Though most tried to avert their nervous eyes, it was clear all were listening as Tom Wright looked down at the butcher's wife and fumed.

"You know, Mrs. Jones, my mam used to tell us children to keep our words of criticism soft and sweet, because there'll come a day when you're gonna have to eat 'em. Your queen bee might just take her hive of talents and find a new home with one of your competitors at twice the pay." His angry eyes ablaze, he stared down at Mrs. Jones.

"Huh?" She crinkled up her nose, entirely lost in the metaphor.

"Good day, Mrs. Jones." Tom tipped his hat, took the two large cuts of beef from Annie, slung them over his shoulder, picked up the block of ice in the other arm, and in long strides took his leave.

❧ · ❧

Annie stood mouth agape as she watched Tom Wright leave the store. No one had ever stood up for her like that before. As the door shut behind him, Annie felt a tap on her shoulder and turned to see the chagrined butcher standing there. He held out a small package of sausage.

"This one here is also for Mr. Wright," he said. "Do you mind taking it out to him?"

Annie's heart beat wildly as she grabbed the package and ran out the door of the shop, hoping to catch him.

There was a warm kindness in his deep brown, captivating eyes when he saw Annie coming with the last of his packages. It sent a wave of warmth through her she had not expected.

"Thank you, Miss Dale." He flashed a dazzling conciliatory smile at her.

As Annie handed him the package, their hands touched, and an involuntary blush flooded across Annie's face. She felt like an idiot standing there being dazzled by him. *Get ahold of yourself, girl*, she thought. She tried to say something witty, but her dry mouth was full of marbles.

He loaded his purchases into the wagon and endeavored to exchange pleasantries, as he sometimes did, but this time he lingered a little longer. She thought he seemed interested in what she had to say.

"How's your mam, Miss Dale?"

"She's fine. Thanks . . . Thank you . . . for asking." She stumbled over her words just a bit.

"And Emma—is she enjoying her work at the dry goods store?"

"Yes. She loves meeting people." Annie gave a weak smile. "She's very social."

"And Grandma?"

"They're all fine," she said in timid voice. "I'm so sorry for you having to . . ."

"No, Miss Dale, *I'm* sorry. She has no right to talk to you that way. I've watched you. You work hard. You're a quick learner, and you anticipate customers' needs. The butcher is very fortunate to have you."

Annie blushed. He hadn't needed to help her, but he did. It made quite an impression on her. "Thank you," she said. Who would have thought she'd find the cranky butcher's wife as footing for conversation with Mr. Wright?

Looking to change the subject, she asked, "You're getting a late start home, aren't you? It's a long drive back to Elsecar."

"It'll be after dark by the time I get home, but I enjoy the ride. It gives me quiet time to think."

"Hmm. Maybe someday it will be possible to drive those nineteen miles home without horses," Annie ran on nervously. "I hear the first train in Europe had its maiden run not far from here—Manchester to Liverpool, I think. Trains will change the face of the entire country someday, don't you think?" She could see her response surprised him and wished she hadn't said it.

"How old are you, Annie?"

"Fifteen," she whispered, wishing she were a bit older.

"Fifteen." He laughed pleasantly. "You're wise beyond your years, young lady!" He looked at her for a moment, which brought an involuntary heat to her face. "Please forgive me for saying it, but I have jars of my mam's honeyberry jam sitting on the shelf at home older than that. Where did you learn about trains?"

Her blush of embarrassment deepened. "I read a lot," she admitted, shyly looking away. "I fear I read far too much for my own good."

"Did Mrs. Jones tell you that?"

Annie felt uncomfortable but exhilarated. It was like he could look right into her thoughts—a bit uncanny, really. She knew he was trying to make her feel comfortable. Trying to coax the words out of her.

Tom continued, thoughtfully. "You seem to be full of surprises. The face you show those around you is unfaithful to who you really are. I'm thinking there are many sides to you, Miss Dale, and most people know very few of them." He paused, clearly puzzling over the discovery. "I love to read. I wish I had more time for it."

"You don't think it's a distasteful waste of time for a girl to have her nose stuck in a book?"

"No, absolutely not!" he laughed. "I suppose that's Mrs. Jones talking again."

His was such contagious laughter, Annie couldn't help but smile in reflection.

"Someday!" Tom was lost for a moment in contemplation. "I would love to just settle in a big stuffy chair on the porch overlooking my own farm and read, read, then read some more from my own library full of books." He looked off into the distance. "Of course, that's just a daydream. But then I think we all need dreams, don't you?"

"I think it's a wonderful dream," Annie gushed, now fully engaged in the conversation.

"Reading adds a depth to your character," he continued, looking at her, serious now. "I started reading as a promise to my mam when I began work in the mine as a boy. It has opened my eyes." He paused to gauge her reaction. "I think you have a wonderful bad habit. And I would strongly encourage more of it," he confessed in earnest.

"You would?" She was pleasantly surprised.

He laughed. "There's a redemptive nature to reading. It educates, illuminates, and enlarges the soul. Books give us both a literal and spiritual way out of the coal mine, and the butcher's shop, if you will . . . and the shortsighted prisons of other people's minds. It can take us anywhere we want to go. Wouldn't you agree?"

Annie nodded vigorously. "Absolutely! Books certainly give us more to talk about than shanks of lamb."

Tom chuckled. "Indeed, they do."

Imagine that, she mused, *a coal miner who reads and doesn't mind—*

"You know, Miss Dale?" he interrupted her thought. "When you drop that guard of yours, let your real self show through, it's clear there is more to you than you let on. You're intelligent, interesting, and you have been blessed with contagious laughter."

Now embarrassed, Annie looked down at her feet, searching for a witty response, but her head was too clouded to find one. She'd said just about enough on the subject already, she decided, and if she said more, she was afraid it might disrupt his apparent good impression of her.

"Miss Dale, I'm sure you hear this so often you're bored of it, but you would be a real asset to any of the businesses around Barnsley."

She crinkled up her nose. "It's not so boring," she smiled, flattered at the compliment from this kind man. "You'd be surprised how boring it ain't."

He smiled with a little laugh at her emerging humor. "You should consider looking around town." He looked thoughtful for a moment. "What do you do over lunch?"

"Oh, I usually go down to the bridge and listen to the river while I read."

"I see." He gave her a playful smile. "And what does the river tell you?"

"Well, Mr. Wright . . . that's between me and the river."

"Very good, Annie! You've bested me," he said as he climbed up on the wagon seat. Then leaning down with another warm, approving smile, he whispered conspiratorially, "Miss Dale, it's certainly been enlightening, and I think you should consider your options." He tipped his hat. "But as you say, it's a long drive home. I better be on my way. Until next time then."

And with that, he snapped the reins and was gone.

My options, she mused as she watched him drive off. "I wish I had some," she murmured, her mouth dry. "And the river tells me, hope as I might, I ain't never gonna find a boy like you."

Chapter 13

 ⤜ • ⤛

Edie hugged Lydia warmly as she welcomed her to their Sunday picnic on a beautiful summer afternoon.

Andrew shook Lydia's hand in both of his. "Thanks for coming, Lydia."

"Look at this, Joey," said Lydia as both looked admiringly at Edie's elaborate spread on a paisley picnic blanket, with matching plates, napkins, and utensils. There were baskets of growler pie, Yorkshire pudding, and other artfully placed delectable delicacies amongst the wildflowers overlooking the River Dearne.

"That's my Sis—the master planner," Joey said, his brows quickly rising and dropping again with a warning glance toward Edie. "Best beware. I'm afraid you're in her sights, Lydia."

Edie ignored her brother. Threading her arm into her husband's, she offered, "Andrew helped. I wanted it to look nice. It was so difficult to convince Thomas to pull away from his work. And I so want you to meet him. He should be here shortly."

"Always plotting!" Joey grinned. "Edie, she's not interested."

"Yet." Edie's eyes glinted mischievously. "Maybe you can tell her *why* she should be interested, Joey," she encouraged. Andrew rolled his eyes at his wife's clearly unwanted persistence.

"Hmm!" Joey sighed. "Why she should be interested in ole Tom?" Joey smiled at Edie's bright eyes, waiting for the further enumeration of his brother's endearing qualities. "My brother Tom is an undefinable mystery. He has . . . a certain surety of character, shall we say? He's brilliant with anything related to machines, thermodynamics, or explosions, and the like. People generally rely on his judgement in technical things, his steady hand, and his calculated approach to solving operational problems." Joey paused dramatically. "But with people? He's not so good. Today, I watched him chastise an arrogant company man who had come down from Woodhouse to 'set some of the young workin' men straight.' Tom brutally embarrassed the poor fellow into stuttering out an incoherent response in front of the boys. These company men ain't used to workin' men having Tom's sharp caustic wit and command of the English language; he's supposed to be an ignorant lout."

"What happened?" Lydia asked with furrowed brows.

"The company official scurried off with his tail between his legs. The young miners being chastised tried hard to thank ole Tom." Joey smiled amusingly. "But Tom was havin' none of it, turned and walked off without engaging them in conversation. Leaving the young miners wondering just what happened."

"Hmm."

"My brother finds it difficult to show appropriate deference to his betters. He often strays from his proper place in the social order of men—embarrassing those he shouldn't. If he's not careful, one day they're gonna set loose Candyman Boo . . ."

"Oh, stop it, Joey," Edie interrupted. "You're scaring' the poor girl. He's not like that, Lydia, really." Edie paused. "Well . . . maybe a little."

Lydia nodded. "I'm familiar with the kind of men of privilege you speak of. I've often had to bear their boorish arrogance during my time at Woodhouse," she said. "I can appreciate your brother wanting to push back. It must be a nice feeling to take those egos down a notch or two. I sometimes wish I were a man and could give them a stomach punch."

Andrew nodded. "Edie's brother stays to himself, his books, and his mining projects. Despite his unsociable nature, his safety improvements and defense of children in the mine have made the miners his biggest admirers, which irritates the company even more."

"He's a recluse," Joey laughed.

"Oh, you two," Edie sighed.

Lydia smiled, thinking she might actually like to meet him.

"He's not your type, and of course, he's not near as handsome as me!" Joey held out his arm. "May I escort you on a walk along the River Dern, Miss Kaye?"

"You may, sir," Lydia laughed and took Joey's arm, much to Edie's dismay.

"We'll be back, Edie," Lydia said.

"Hmm . . . maybe?" Joey added, as he sauntered off arm in arm with Lydia.

⊁ • ⊰

Tom strode up to the picnic in haste, reaching out to give Edie a hug. "I'm so sorry I'm late." He shook Andrew's hand. "I got held up at the butcher's shop in Barnsley, then had to help Mam with Papa."

"He's looking pretty feeble these days, isn't he?" Edie responded soberly.

"He is." Tom nodded. He looked over the elaborate picnic spread Edie had prepared. "Everything looks wonderful."

Edie beamed at her older brother's compliment. "Thank you, Thomas!" Then turning to Lydia and Joey, who were lost in conversation as they strolled up the gentle slope from the river, she invited, "I'd like you to meet someone."

Edie tugged Tom's elbow in their direction and gestured to Lydia. "Lydia," she called out, "this is my big brother, Thomas."

Lydia looked up from under her parasol, her dancing eyes and engaging smile quickly vanishing when she saw Tom.

Tom, who had not yet released his attention from Edie, slowly turned toward her new friend. His brows lifted in surprise when he saw her. Then for a long moment without a word, he stood peering up at her from dark squinted eyes under lowered brows.

⊰ • ⊱

Lydia saw the recognition register in the squint of Tom's disarming eyes. Despite feeling self-conscious under his penetrating stare, she said nothing. Her eyes narrowed, and with tight lips, she stared right back at him, hoping she might get through the ordeal without appearing flustered. But when his intense gaze took on an unabashed hint of a smile, her heart skipped a beat. The remembrance of that breathless moment in the forest flooded into her mind. As her flustered eyes took in the disturbing expression of this tall, lean, and exasperating man, her hot blood brought an involuntary blush of pink to her face. Irritated with herself, she wondered just how he could do that without a word said.

In recognition of her distress, his smile spread wider across his attractive face.

But there was no uncomfortable moment from which the beautiful Lydia Kaye could not quickly recover. With a disarming sparkle in her emerald eyes she delivered a bright smile of her own that could melt cold, hard steel and immediately turned the tables.

Edie turned to Andrew, confusion etched across her face, brows forced together in a questioning frown. But Andrew, looking equally puzzled and wide-eyed, could only shrug his shoulders.

"Lydia," Tom nodded in acknowledgment, without taking his knowing eyes off her. It was enough to confirm his understanding of the unspoken message that had passed between them. "It's nice to meet you at last." He smiled.

Edie stared at the two of them, clearly irritated that she was not invited into their unspoken conversation.

"Thomas," Lydia nodded in terse, reciprocal acknowledgment, struggling mightily to hold his intense gaze. This time she was intent upon not giving him the satisfaction of making her look away. His irritatingly contagious smile spread still wider across his face. His demeanor gave the impression of a stern calmness, a clear sense of intelligence, and a hardness suggesting he was the kind of man not easily drawn into casual conversation. His dark, restless eyes with flecks of luminescent green seemed to hint of a secret within. One he would never reveal. It was like he knew something of her thoughts, which of course, he did.

She wanted to reach up and wipe that smile right off his face, but in the company of his family, she only offered a charming smile of her own in return. Try as she might, she could not stop her face from betraying the visceral attraction she had at seeing him. Lydia was clearly interested in this man, despite her reticence to admit it, and he in her.

"Why do I get the distinct impression that there is a private conversation going on here?" said Edie, exasperated at missing out.

Both Tom and Lydia held each other's confused eyes without responding to Edie. The secret of their enticing few moments shared before daybreak on that never-to-be-forgotten morning would remain undisclosed.

"Look, you two," Edie declared. "This is starting to look a bit indecent. I think Andrew and I will take a walk along the Dearne and leave you two alone to finish your . . . whatever this is."

"Okay," muttered Tom distractedly, continuing to look at Lydia. "We'll stay here and talk a while, if that's okay with you, Lydia?" Tom nodded toward her for confirmation. She smiled agreeably.

"Okay," Edie gave a resigning smile to Andrew. "Well then . . . we'll be back later. Come on, Joey. You're coming with us."

⤙ • ⤚

Lydia was intrigued by this tall, mysterious man, handsome, with an engaging smile who seemed to be chasing his inner demons. Maybe he could be a friend. Maybe . . .

"How is your father feeling?" Lydia asked.

Tom frowned. "Not well, I'm afraid. But thank you for asking."

"Edie says you almost single-handedly kept the family together when he came down with black lung."

"Oh, I wouldn't believe everything Edie tells you. She can be a bit melodramatic at times. You know how kid sisters are."

"Actually, I don't. I never had a sister. After my papa sold me to Woodhouse, I never had any brothers either. Most of my growing up was done in service to the Fitzwilliam family."

"Sorry, I didn't know." Tom gave her a long inquisitive look. "It was touch and go for a while after Papa got sick. We all had to pitch in."

"Is that when you got into the business of poaching?"

"Poaching?" he replied with a frown. "Such a harsh word. I prefer to think of it as 'creatively acquiring a decent meal to keep the Grim Reaper a safe distance from my family.'"

Lydia laughed. "It's a capital offense, I hear. And Lord Fitzwilliam is not a forgiving man." She looked at him with apprehension. "It seems a dangerous hobby."

Tom didn't answer for a long time. "So, what else has Edie told you?"

"Quite a lot, actually. Although apparently not everything." Lydia smiled. "She told me about the devastating loss of your grand, and your boyhood friends."

"Hmm. Edie can be injudicious at times." He looked at Lydia for a long moment without expression. "I learned the art of acquiring food for the poor from my grand. Before he was blown to smithereens in Lord Fitz's mine."

"I'm so sorry." She paused to gauge his reaction. "Were you close?"

"Yeah, we were close."

"It must have been difficult. I always wished I'd had family to be close too. It was hard not to be angry. I found I needed others just to keep my sanity, to avoid becoming bitter. I suppose we must live with the deck of cards we've been dealt in life."

"Hmm!"

"They say fear is the enemy of happiness. The only way to take the heartbreak out of death"—she paused taking note of his skeptical reaction—"is to take the blessings out of life. We need each other, don't you think?"

He smiled at Lydia thoughtfully. "I suppose I could make an exception in your case."

"Why? Because I'm beautiful?"

"You are that," he said. "But then, life isn't so bad living without the complications of other people."

"Really?" she challenged. "I can see you have a lot of work to do, Mr. Wright . . . to get your thinking right."

"Are you applying for the job?"

"Don't get your hopes up," she eyed him shrewdly.

Tom soberly looked at Lydia for a long, disconcerting moment. She could see his mind working. He was clearly confused, feeling off-balance. After a long moment of indecision, he asked, "Could we sit for a while and talk?"

She nodded her consent. Tom reached over, putting his hand under her elbow as they walked over the uneven ground to sit on the summer grass below a great spreading oak. His other hand rested on the small of her back. At his touch her breath caught, and her complexion flushed at the unsettling sensation, with the warm glow of beguiling confusion. Fielding these feelings would be complicated, making her plans to remain aloof more difficult.

They spoke of many things that afternoon—of Grand, of Tigger, of a little girl growing up in the foreboding Woodhouse Mansion. They shared their mutual appreciation of the mysteries of nature, the creatures of the forest. He told her of his midnight hunting ventures under the stars, and she shared her memories of early morning walks. She recalled the mesmerizing beauty and quiet precision of the enchanting great grey heron, a vision of perfect control on silver-grey wings, who glided over the estuary, landing deftly on the water's edge. In pursuit of her prey, with patience, skill and precision, this master magician of stealth stepped slowly forward, a hunter stalking its prey in the shallows among water lilies so green, they seemed to glow with tiny blossoms of inflorescence at daybreak. Then, like a lightning strike, her beak hit the water, spearing her unsuspecting prey with deadly force.

"Sounds like she can be deadly to her prey," Tom responded.

Lydia pondered the man who sat beside her, plagued by his demons.

"I suppose this is another of my impertinent questions, but I am dying to know. What happened at the pit manager's office after your run-in with Boo Black?"

"Hmm." Tom smiled. "Joey, or Edie?"

"Both."

"I should have guessed."

Tom took a long moment to collect his thoughts. "Well," he began. "The morning after my infamous debacle, I sat nervous in the pit manager's front office. I was a terrified boy with a head full of mush wondering just how my life was going to end. The pit manager's assistant was noisily shuffling papers around on his desk while I wrung my hands, looking for a way to escape. I prayed for a reprimand, hoped for a beating, but expected worse. My whole body had broken out in a rash. I was perspiring so badly, it was embarrassing." Tom paused, chuckling at the memory.

Lydia encouraged him on with a smile. "I can imagine."

"I thought to myself, *So many mouths to feed. What an idiot.*"

Lydia looked on anxiously.

"Finally, the pit manager called me in.

"'Young man,' said he, 'I've been askin' around, and it seems you're pretty handy with mechanical things. D'ya think ya can repair the bilge pump?'"

"What?" Lydia interjected.

"Exactly. I fumbled with my hat in my hands. 'Pardon me sir?' My eyes must have been wide in confusion. I was so focused on hearing what punishment was to be meted out, I wasn't at all prepared for that question."

"'Can ya fix the pump or not?'

"'Uh . . . sure. Yes sir.' I says. 'I can do it, sir.'"

"'Well, get to it, then,'" Tom imitated the pit manager.

Lydia threw her hands up in relief. "No beatings. No clubs. Not even a tongue-lashing?"

"Nope! Of course, I had no idea how to fix a bilge pump," he laughed, which brought a smile to Lydia. "But my life wasn't over, and for that I was as happy as a pig in mud. Before I finished with that pump, I had hundreds of parts spread out all over the floor. Piece by piece, I put them all back together, fixin' each broken part along the way. When I flicked on the switch with my fingers crossed, it hummed like a prayer to sweet Jesus, belts turning, pistons pumping. And just like that, I was on my way to a new future."

"What a great story," Lydia exclaimed, thinking Tom looked handsome telling it.

"Yup. Ole Supe seemed to think I was some kinda creative genius repairin' stuff. From that day on, I was constantly being asked to make machinery repairs, come up with inventive improvements to all kinds of equipment—pumps, pulleys, winches, steam engines, fans, heating systems, vanes, impellers, ventilators at the iron works foundry. It seemed that anything that could fail in the operation of the mine, did. And they'd call me to come fix it."

"So, you're pretty good with technical things, huh?" Lydia said.

"A genius, they say." Tom winked with a laugh so contagious, Lydia couldn't help but join in. "I'm good with other things too."

"You think so, huh?"

Ignoring the undercurrent of meaning, he spoke with a deadpan face. "I found the study of science and technology intriguing. It gave me the tools to solve problems. But what I liked—*really* liked—was connecting my fingers to the engines of my mind. I liked fixing and building things."

"Not my cup of tea, but fascinating!" Lydia said with a smile.

"At first, when the miners saw me comin' shoeless, my filthy hair sticking up in every direction, baggy pants, pockets everywhere," he remembered with a contagious smile, "my homemade wooden tool-box in tow, they stared at me with a skeptical, knowing smile."

≁ • ≺

Tom remembered the long-ago time and what the miners had said of him. His mind seemed to work differently than others. "The boy wonder," they'd called him, who kept to himself but always seemed able to shape his mind around the problem and turn disorder into order. Then, mysteriously, the obstacle impeding the operation of the machine revealed itself, and just like that, he'd fixed the problem. Before long, Tom had become the resident expert on anything mechanical, technical, and prone to explode. But, much to the irritation of many in company management, costly safety items seemed frequently to sneak into his repairs. The miners quietly acknowledged these side benefits, and the company not so quietly complained about the cost. But Tom had saved so much time and money, not to mention increased production, that most in management grudgingly turned a blind eye to their now well-paid young technician.

Despite his disdainful personality and unsociable nature, the miners at the Silkstone came to respect the boy with the answers and backhanded insults to the educated company men.

"The company started farming me out to the other mines along the Pennine coal seam. Eventually, Lord Fitzwilliams offered me a job to not only work with their mines, but their ironworks foundry too," Tom shared. "So, the family picked up and moved a few miles down the road, and just recently bought a little old farmhouse at the base of the Pennines hills just outside of the Wentworth Wood-house Estate forest. It's very small, but better than the one-room shanty and close to Elsecar. Not far from where I saw you for the first time."

"Hmm!" she slyly smirked, knitting her brows together. "I vaguely recall that meeting in the forest." She caught his gaze and they both smiled. "Joey told me you were the youngest Silkstone & Elsecar Coalowners Company engineering manager ever."

"Did he? I'm not so young anymore."

He continued to puzzle over the beauty of this woman with a quiet grace and elegance. There was a confidence in her demeanor, a surety in her character, an unassailable dignity in the way she carried herself. He was sure her bearing alone must force men to show her respect, and most to hope for more.

Tom looked at her for several moments. He was not used to being out of control, and now found himself considering breaching the wall he had carefully constructed to protect the solemnity of his solitude. "Would you mind if I called upon you?"

"When I first saw you, all I saw was this sullen man. Now I see the boy within," Lydia said softy. She reached out and touched his cheek. "I think I might like that. But I have to tell you, I'm not inclined to share you with your demons."

⊱ • ⊰

Having introduced the state-of-the-art exhaust fan that recycled cool, fresh air into the dining and reception hall, Tom had received repeated invitations to the Woodhouse Estate. The talented technician was conscripted to improve, redesign, and even invent several new and imaginative systems around the massive estate—creative conveyor belts, a boiler room redesign invented in America, a water pressure system, lift stations, and other improvements to the 310-room, 255,000-square-foot mansion. There always seemed to be some system to redesign, install, or conceive, and money was no object. But Tom's most outrageous suggestion was a flush chamber pot. Of course, Her Ladyship was incensed that a mere servant should talk of such personal things. The idea that she might give up her chambermaid standing by during the night to wipe her *derriére* with silk linen and immediately empty the chamber pot was of course absurd. "The man should be whipped for his insolence," she told her husband upon hearing the idea.

In the end Tom was given permission to install his Wedgwood-Crapper flush toilet in one of the far distant fifty-seven bathrooms as an experiment. Its installation was a curiosity to all in the great house. Both skeptical servants and curious residents alike stopped by to look in on the work and eventually give it a test run. They couldn't imagine keeping smelly night soil inside the mansion would ever be condoned by the Fitzwilliam family. But the miraculous invention was so successful that Tom gained a new reverence from the staff and became a regular visitor to the great mansion, installing these and other modern conveniences in all the bathrooms.

Tom's latest project included the installation of an ingenious piston-operated dumbwaiter, designed to quickly deliver food and wines from the kitchen and cellar to the dining room, and just as quickly clear away dishes and anything else that needed to go. He was in the middle of installing the third of three such systems for an upcoming dinner party of unthinkable extravagance, when a nervous Candyman Boo Black was escorted into the nearby service reception hall.

Boo paced back and forth, muttering to himself. Tom, having heard rumors, knew Boo was in for a thrashing by Lord Fitz.

"Whatta ya lookin' at?" Boo spat out with an angry frown when he caught sight of Tom. "I'll be wipin' that smile off your face."

Looking in Boo Black's direction, Tom continued working without responding in word or expression.

After a time, Lord Fitzwilliam entered the room. His curt, enigmatic nod to Boo Black did nothing to warm Lord Fitzwilliam's dark, foreboding eyes, nor take the sharp edge off his condescending demeanor. Ignoring Tom's presence, the nobleman began to berate the Candyman, who stood humble and silent, hat in hand, his face and scalp under sparce, matted hair sweating profusely as he looked down at his feet.

"I expect those in my employ to give their full and complete allegiance to my family and the wishes of my company," Lord Fitzwilliam resolutely began, the grandeur of the Woodhouse Mansion setting further fortifying his demands. "I expect you to enforce those standards on everyone in my village. No more of this nonsense."

"Yessir," Boo responded without looking up.

"It is your role to make sure they cease this incessant whining about the children, safety in the mines, and food for those who don't earn it. It needs to stop." As tall as Boo, Lord Fitzwilliam paused just inches from Boo's downcast face. "It's been this way for two hundred years and, by God, it will remain this way long after you're gone. Which might not be long if you don't get to it, man," he directed coldly. "Do I make myself clear?"

"Yessir, Your Lordship."

"We must all understand just where our duty lies. Mine is to rule. Yours, Mr. Black, together with my servants, miners, and their families, is to serve *my* family; to bend to my will; to follow the direction I see fit to give; to do your duty under the dictates of my good graces, and, for heaven's sake, to do it without complaint. Is that clear, you fool?"

"Yessir!"

"There is no room for dissidents here. No malcontents. I won't tolerate it," he said coldly to the cowed brute of a man. "My meaning is not confusing, is it, Mr. Black?"

"No sir. I understands." Boo humbly bobbed his profusely sweating, nearly bald head in acknowledgment. "I get your meanin', sir. And ya can count on me to see to it, Your Lordship!"

Lord Fitzwilliam twisted his mouth into a sneer, pulling in a deep breath and holding it in frustration. "I will not waste my fortune on these frivolous luxuries for those who don't merit it. This is your lot, the life you were born into. You will do as I say, or there'll be hell to pay. Now get to it, man."

With that, Lord Fitzwilliam turned and walked away.

His rapid-fire insults had been driven home with little regard to the terrifying retribution that would be meted out by Candyman Boo Black on the families in the village. The exact details of that unseemly affair were not His Lordship's concern. As far as Lord Fitzwilliam was concerned, Tom had no eyes to see, nor ears to hear, nor did he care if he had a mind to comprehend. But it was clear Candyman Boo Black knew otherwise. Thoroughly humiliated, Boo fumed as he turned to leave. Vengeance glared from his lonely eye in Tom's direction.

Chapter 14

♄ · ♄

TOM DID CALL ON LYDIA that following Sunday, and the Sunday after that. The weeks rolled into months. For the first time in a long time, he had been able to draw in a deep breath of fresh air. It was as if a heavy weight had been lifted.

The months rolled into a year, then another year and another. Tom came to believe the day he met Lydia Kaye had been the luckiest day of his life. He came to recognize that finding the right companion to spend the rest of his life with was the closest he would ever come to determining his own destiny.

Both Tom and Lydia entered into a golden warmth of understanding that they would marry. They had grown together in heart and mind. In his old life Tom had been aloof from others, rarely spending time away from his work, study, and seldom in the company of anyone, even family. But now, he shared goals with Lydia, planned for their future together, and lived for those precious Sundays when they would spend the whole day together. It felt good to be alive and in love, and there was comfort in knowing what his tomorrows would bring.

On a beautiful spring Sunday morning, Lydia met Tom at the door of the chapel. He was late, of course. The church service had already begun when they slipped in to sit beside Martha and the rest of his family. The split-log pews were completely filled. Alcoves cut into the rough-hewn timber walls were lit with candles, spreading pools of light to mingle with the shadowed morning sunshine.

At the front of the chapel stood Bishop Walker, whose quiet, soothing voice reached into every corner of the chapel with a message that drew in the congregation. Educated in London, Bishop Walker was a tall, slender, middle-aged man of the cloth with kind eyes and a calm, thoughtful voice. For most of twenty years, Bishop Walker had taken a special interest in Martha and her family. Martha, Edie, and most recently Lydia, attended his services religiously each Sunday morning, and of late, Thomas had joined them.

While Lydia seemed to feel a closeness with her many new friends in the congregation, Tom, who did not share her religious conviction, was there to be with Lydia. She wanted to be in church, so he came. In the hush of this Sunday morning service, Tom felt a warmth course through his veins as he held Lydia's small, white-gloved hand folded in her lap. He was overwhelmed by her delicate beauty. Her thick auburn hair tied back in a refined twist, in stark contrast to her clear, creamy complexion.

Lydia's features made a striking portrait. Her dignified manner had been imposed on her by the strict discipline of her years in service at Woodhouse Mansion, but her fiery lust for life and her turbulent willfulness were her own. If anything could reflect the source of her enchanting charm, Tom thought it had to be her emerald-green eyes dancing with mirth. They lit up her mischievous smile and held him captive. Tom knew Lydia was well aware of the effect her feminine wiles had on him, and she wasn't a bit shy in using them to her advantage. He also knew that although she did her best not to show it, she absolutely adored him.

In the careless ease of youth, they shared soft laughter in the aisle after church service on this Sunday morning. As quickly as they could, they headed out to breathe in the fresh air of springtime on a long walk down the narrow country roads, defined by hedge-rows thick enough to stop a runaway coal cart. The hills around them were covered in blazes of springtime wildflowers—daffodils, daisies, and splashes of primrose amongst the groves of alder, birch, sycamore, and here and there, the great oak. These picturesque spring afternoons with just the two of them were worth their weight in gold.

Between Tom's endless weeks at the mine, and Lydia's long hours in the match factory, they had little time for themselves. After leaving service at Woodhouse, Lydia spent fourteen-hour workdays in far-off Sheffield, bent over the phosphate bins dipping match sticks at the match factory by day. Her nights were spent in a dormitory full of girls. It was only on Sundays Lydia and Tom could be together, and sometimes not even then.

As they walked the secluded pathways, scattered lazy clouds driven by distant sea breezes drifted east in a brilliant blue sky. Lydia and Tom walked, talking playfully, their laughter echoing around them.

"Hurry, Tommy! You're so slow! Am I going to have to always wait for you, Pokey?" Lydia laughed, running out in front of him.

"You'll be a sorry girl when I catch you," Tom called after her.

"I would apologize for bruising your fragile pride, sir," Lydia teased, running still farther ahead. "But I'm simply too excited about the prospect of beating you yet again . . . Pokey!"

With a quick burst of speed, Tom caught up to her and grabbed her from behind.

"You'd best apologize to me, young lady." He pulled her into his chest and wrapped his strong arms around her. Her hands rested on his chest as she laughed.

"Unfortunately, apologizing does not come easily to me," she demurred, offering her most warm and alluring smile.

Tom grinned. "Admit it. You think I'm handsome."

"I don't know, Tommy. A woman has just gotta have standards." She frowned wryly, pulling her brows together in consideration. "I'm so sorry I hurt your tender feelings," Lydia began coyly mocking him.

In a moment of laughter, he loosened his grip on her, and she broke free from his grasp to run out ahead yet again.

"That's your apology, Match Girl?" Tom called out.

"Even in your wildest fantasies, you couldn't catch me, Pokey," she teased. "And don't call me 'Match Girl.' You're speakin' 'ta a lady here—the best woman ya'll ever know!"

Lydia was panting and out of breath when she gave up running. "I'm just taking pity on you, Tommy," she said, leaning over, putting her hands on her chest. She began to cough, then coughed again and couldn't stop.

Tom gathered her up into his arms. "Are you all right?" He helped steady her. "You still have that awful cough?"

She coughed again. She was having trouble catching her breath, then another coughing attack came. "I'm wise to your sympathies, mister. You just want to hold me 'cause you think I'm pretty!" Giving him her best coquettish smile, she added, "You want to kiss me, don't you?" She poked her finger into his chest to accentuate her point.

He stepped back for a moment, leaving his hand holding onto her arm. "Are you sure you're all right? I'm concerned about that cough."

Lydia spun loose once more, batting her eyelashes. "Personally, Tommy? I think a boy who tries to use his unwarranted concern as an excuse to kiss a pretty girl does not deserve her. You need to concentrate on wooing me if you expect a kiss."

Tom wasn't fooled. He could see the worry in her eyes. "Lydia," his brows knit together. "It's been how many weeks now?"

Her smile drooped a little. "It always has to get a little worse before it gets better. But I'm fine, really. *Really*, Tommy. Let's enjoy our day and not worry about my cold, all right?"

Tom sighed, accepting that she didn't want him to push her on the subject. They walked together hand in hand, smiling at each other the last half mile without another word spoken. Both were eager to get to their favorite picnic meadow by the creek.

"There it is!" Lydia pointed ahead to a small meadow with a grove of giant oaks, spreading sycamores, and languid magnolias canopying the meandering brook running through the meadow. "I think God made this place just for our Sunday afternoons," Lydia sighed, sending an inviting smile Tom's way, further warming his heart.

⤛ · ⤜

"Tell me how it's going to be, Tommy," Lydia begged as she lay back on the picnic blanket. "I love to see your face light up when you talk about our future. Tell me about our land, the little farm where I'm gonna raise my boys, and spoil my little girl so terribly."

"Listen, Match Girl. We don't need no damn little nits prattling about under our feet."

"Oh, Tommy. When you're losing an argument, you don't strengthen it with profanity. I can see I still have some work to do to make you into the man you almost are. The man I know you can become," Lydia sighed playfully.

Tom ignored her censure, his mind off on another daydream. "In America, you are only limited by your own ambition, imagination, and your willingness to work for your dreams." He paused, his visionary eyes alight with excitement. "People aren't owned by the company store or subject to some tyrannical aristocratic government. Folks are free to pursue the life of their own choosing." He let his thoughts run free. "Can you imagine it—to live your life with no one deciding what you must do?"

Tom lay back on the blanket and put his hands behind his head. He stared up into the billowy clouds floating across the deep blue sky, allowing this dazzling prospect to capture his imagination.

"I can't wait," Lydia sighed. She slid in beside him and made herself comfortable, her head cradled in the crook of his arm, her hand laid atop his strong chest. "Tell me more."

Tom pulled in a breath to fill his lungs with the smell of springtime. "Out west, they have land as far as the eye can see. It's just there for the taking. I hear they've got gold nuggets in streams that a man can pick up when he has a mind to." Lydia was caught up in the longing in Tom's eyes. "I want land, acres of it, with a crystal-clear stream running through it. Land is the only thing that lasts. It's not just the symbol of wealth, land *is* wealth."

"I suppose it is."

"We'll grow corn, alfalfa, maybe even have an orchard. And pigs . . . yes, I think I'd like to have pigs. Big, fat, noisy ones that roll around in the mud."

"Pigs!" Lydia's brows pulled together in a frown and she crinkled her nose. "Seriously, Tommy?"

"I want to go fishing in my own brook," he said, too distracted by his dream to respond to her complaint. "I want to catch a trout with a pole." Tom had been captivated by the wonderful florescent colors of trout in the forbidden forest, their muscular, torsional beauty and iridescent patterns glistening in the sun, like an imprinted code of life left by God's own hand. They had seemed so close, with their white-ended fins wimpling in the water. He'd thought he could almost reach out and touch them. "Can you imagine?"

Lydia couldn't help but catch his enthusiasm. "I could send you out to catch our evening dinner while I sit by the fire playing with our little one and eating chocolates."

"Trout for dinner, just as nice as you please. I've heard tell they have streams that are so clear, you can even drink from 'em. Clear as icicles. None of that black water thick with coal tailings."

"Livin' off the land just 'cause we have a mind to," Lydia responded, gazing into the clouds drifting overhead. "Aye, it must be heaven."

Tom chuckled. "With a bit of luck, we'll make our own heaven on earth. Who needs an afterlife? I'm strong enough to achieve my own goals in this life without making weak excuses for failure. Heaven is just a place for the feeble-minded to deposit their troubles."

"Oh, Tommy. Don't be sacrilegious." She frowned. "Really, Mr. Wright. You may be a handsome brute, but I still have some work to do with your misguided thinkin'."

Tom shrugged. "Besides, I'm not sure I'd want to make a commitment to heaven . . . or to hell. I got friends in both places."

Lydia frowned. "Well, aren't you the cynical one, Tommy Wright! Could there be anything more American than faith in the Almighty? Spending your childhood down in the coal mine must have really played hob with . . ." She paused. "But not to worry," she said. "It's my job to fix you. And I will. Just leave it to me, and in no time, we'll have you pulled back from the brink of eternal damnation."

"You're too much, Lydia."

Lydia handed him a bowl of Yorkshire pudding. "I think I'll have four, five, six children. Maybe even more! They'll play in the meadows and run down to that stream of yours. No more alleyways in front of old Mrs. Straycoff's shanty. No mine shafts. Just meadows and brooks to play in."

Tom, shaking his head, looked askance at this beautiful woman.

She gave him a sly smile. "I'm the best woman you'll ever have, Tommy." Cutting a slice of growler pie for each of them, she took a big bite of the luscious delight. "Mmm, that's good! Maybe we'll have pie and trifles after every meal when we go to America. Maybe I'll just get as big as a house, happy as can be, rolling around like a rolling pin with all my children!"

"I don't think so, Match Girl. Look at you. I swear you're getting skinnier by the day." He sighed. "What am I gonna do with you?"

"You're gonna love me and give me what I want. You're gonna make me a mam!" she declared in a breathy giggle, a sparkle ablaze in her green eyes. "You will be a wonderful papa. And imagine—after all our babes have gone to bed, we can sit out on our porch on a summer night with real stars overhead." She snuggled up tight, her head on his chest, holding his hand in hers. "Oh, it's gonna be grand."

"You'll never give up, will you?"

"I can't," she said, a twinkle in her eye. "I gotta make you into a better man. You know in the end, you'll give me what I need. And that's why I love you. Tigger would be proud of you, Tommy me boy."

Tom hugged her. "You're better than any afterlife."

She tenderly ran her fingertips over the boyhood scars on the back of his callused hands, tracing his memories of heartbreaking loss. Her heart ached for him, so much devastation left in the wake of the tragedies Tom had lived through. "My poor, beautiful man," she whispered. "We have a lot of work to do, you and I." She sighed, holding his hand in hers. "But these determined hands reveal the promise that every hurt can be healed. Every dream can be reached."

He smiled and closed his eyes. "We'll do it together, my dear."

"How much longer until we can make our dreams come true?" she asked solemnly.

For a long moment, Tom didn't answer. He had told her they needed to save money and wait until the war in America had ended. But what had concerned him most was leaving his family to fend for themselves. Until recently, that burden had landed squarely on his shoulders. "We need to go, don't we?"

"We do," she sighed. Then lifting her head off his shoulder, she gazed up at him with a frown. "And you need to properly ask for my hand in marriage." She added in mock seriousness, "Who knows? If you wait much longer, I might get a better offer." She lay back down on the picnic blanket staring up at the paintbrush of sky.

Tom marveled at Lydia lying there. Her eyes danced, dreaming about their future. He was hopelessly under her spell. For better or worse, he was destined to spend the rest of his life in love with this woman.

"Pander to me, Tommy," she teased playfully. "If it's better than last week's, there may be a little something in it for you—what was it you said?" She crinkled up her nose, trying to remember. "Oh, yes. 'I need you like a heart needs a beat.'" She laughed at his creative endeavors. "I so love it when you pander."

Tom grinned. "This week, rather than a clever phrase, I thought I might give you a gift instead." He reached into his knapsack, pulled out an envelope, and handed it to her.

"Oh, Tommy. I do so love surprises!" She propped herself up on one elbow, holding the envelope in front of her as she looked at him, puzzled. "Give me a hint, Tommy." She held it up to the sky.

"Joey got his promotion to office staff at the mine last week. It comes with more pay."

"Good for him." She looked at Tom, perplexed. "What kind of a hint is that?"

"The Civil War is almost over, and the good guys are winning . . . And we just paid off my parents' farmhouse."

She frowned, but then her discerning eyes lit up in recognition of the accumulated significance of these facts. "Oh, Tommy, don't tease me!" Her hands trembled as she fumbled to tear open the envelope.

She unfolded the letter inside and read aloud, her eyes open wide. "*Reserved: two tickets on the* Brigantine Constellation *from Liverpool, England, to New York City, New York. Payment due by June 1, 1865.*"

"Really, Tommy?" Lydia sat up and covered her mouth with her free hand, laughing out loud. She threw her arms around him. "We're goin' to America?"

"I think it's about time, don't you?"

"Oh, Tommy, you've done it!" She fell back on the blanket and waved the letter back and forth, kicking her feet in the air. "We're goin'! We're goin'! We're goin' to America!" Then, pausing with an enormous smile, she leapt to her knees, nudged him with her shoulder, threw her arms around his neck, and kissed him. "We are going to America, Tommy me boy!"

"And I suppose if we're leaving in two months, we should get married first, don't you think?" He paused with a mischievous smile. "That's if you'll have me?"

"This is definitely better than last week's, Tommy," she said. Slowly, tears came to her sober eyes as the implications wound their way to her heart. "I love you so, Tommy Wright. I'll be a good wife!"

"You better!" Tom chuckled. "'Cause you've got me entirely vexed. I'm no longer accountable for my actions."

"You're unstoppable. It's what I admire most about you." She looked at him with an illuminating sparkle in her emerald-green eyes. Her ardent smile gave way to a warm glow that lit up her face like a bright light in the night.

"I swear, your smile could bring December daffodils into bloom," Tom murmured.

Lydia rolled out from under his arm and pulled him down to her, the overwhelming loveliness of her body touching the length of his. She whispered in his ear, "I suppose you'll be wantin' your reward now?"

He pulled back to see a knowing smile light up her face. He marveled at her beauty. She brought forth in him the wild emotions of springtime, warming him to his very core from which sprung an almost unquenchable ache for her.

"I can see carnal thoughts in your eyes." She paused to make him suffer, then added, "You know, Tommy. You're a naughty boy. And I do so love it when I'm the reason." She flashed a playful frown, then pulling him to her, she kissed him. "Never underestimate the healing power of a good ravishing."

"Stop talking." He pulled her slim body close and felt the swift thudding of her heart as she melted into him. Passion rose within him.

Over the next dizzying few minutes, Tom set aside the memory of the bitter years in the mine, the loss of those he loved, and all his hopes and dreams for the future. He thought only of this passionate moment intertwined with the woman for whom he held a deep and abiding love. Lydia Kaye was more enticing than any foreign land, more exciting than any adventure his mind could conjure, and more alluring than the most beautiful sky over a pristine meadow. For a timeless moment, they held each other in their private meadow, forgetting about everything and everyone, as if this were the last carefree moment they would ever spend together.

Chapter 15

✦ · ✦

ANNIE HAD SEEMED TO AT least try to follow the admonition given most young girls to "be seen and not heard" in her early conversations in front of the butcher's shop with Thomas. He had become like an older brother to her, sometimes offering encouragement in her skirmishes with adolescence, sometimes advice in her struggles to help her define who she was. He mentored her in things of the world and tried to help her sort out her plans for the future.

As she forgot her shyness, she became an unstoppable fountain of insights, and sometimes Tom wasn't sure just who was actually receiving the most profound advice. She seemed to have opinions on a whole range of subjects—books, politics, world affairs, life in America. When she rose above her insecurities, their conversations often took flight, sometimes going on for a half hour or more. Her enthusiasm often made him smile in appreciation of her intellect and uncanny insight. He never knew what direction they might end up heading, or what surprising new thing he was about to learn from the intelligent, well-read Miss Dale. Some of those discussions were downright captivating. Her insights were wise well beyond her years, and oh, how she could make him laugh, sometimes when he most needed it.

One afternoon on one of his frequent trips to Barnsley, she asked about his brother George. Tom explained that George had lost his wife in childbirth at nineteen, and a month later, the baby died as well. "George was so upset," Tom shared. "He packed up everything and headed to America."

"Oh, your poor brother."

"George was always the curious one . . . the adventurous one," Tom sighed. "He began work at the *Chicago Tribune* just before the start of the American Civil War, only to disappear into the perilous frontier two years later. I miss him. We all do."

"I'm so sorry, Mr. Wright."

Annie seemed upset at the turn of events. It touched Tom. "George is the kind of brother who always wants to be at the center of things," he said, nostalgic. "He thinks every opportunity for adventure has an expiration date and is afraid he just might miss it if he doesn't rush into the midst of the action. He often wrote about America's struggle. Abraham Lincoln—"

"The growing pains of a new nation," she interrupted, caught up in the fascination of the subject. "Thousands dying in that war. Slavery. It's changed the way of life for the South, hasn't it? A way of life that we in England helped to foster." Annie caught herself. "I'm so sorry for interrupting you, Mr. Wright—Nanny says I could learn a lot if I can just keep my mouth shut long enough to listen."

Tom laughed. "That's quite all right. But you're right about our country being complicit."

�™ · ☙

Two weeks later, when they were again loading his wagon in front of the butcher's shop, Annie picked up the conversation where she had left off, in classic Annie style. "Mr. Wright, I was thinking," Annie

began, "about George and President Lincoln. I read up on it a bit—
Mr. Lincoln, I mean. His efforts to amend the Constitution and finish
what America's Founding Fathers started a century ago—intriguing."

"You don't miss much, do you, Miss Dale?"

"I absolutely love Abraham Lincoln."

"George was captivated by Lincoln too."

"Born on the frontier," she said, "his rise to power, his obsession
with preserving the union of states. All very inspiring."

"Do you think Lincoln was inspired?"

"Hmm!" she smiled. "He grew up in a poor family with a strong,
determined mam who gave him a homespun education." She stopped
for a moment. "Not much different than you and your family."

"It may be stretching it a bit, but I suppose you're right. There are
indeed some similarities in our upbringing."

"Where else in the world could a man like Abraham Lincoln suc-
ceed but in America?" she added. "It's a place where you don't have
to be born an aristocrat to be successful. A man can begin as a lowly
frontiersman and end up President of the United States solely on the
merits of his oratory skills, intelligence, drive, and determination."

"It's astonishing," he said. "Could you imagine a self-educated
commoner the likes of Abraham Lincoln as prime minister of
England?" He laughed in derision.

"Or a King of England who was born on a poor frontier farm?"
she added. "Only in America." Annie joined Thomas in laughter. "I
envy you. When will you go?"

"Soon." He paused considering. "Once Lydia and I are married
we'll be following George across the ocean to seek our fortune!"

"And become president?" she laughed.

"That's not in our plans."

"Congratulations, by the way. She is so pretty." Annie looked at Thomas, self-conscious of her own plain appearance. "I'd love to go to America someday." Tom could see the wistful look in her eyes. "I hope you're not too late in the season. I understand it's prudent to avoid travel across the Atlantic in the spring and summer on account of the tempests, heavy seas, and even hurricanes in late summer."

"I didn't know that about hurricanes," said Tom. Her knowledge of so many differing things always seemed to surprise him.

Annie's brows came together in concern, carefully considering a question. "Do you mind if I ask what's happened to your brother George?"

Thomas paused for a long moment.

"We don't know. We just lost contact with him," he sighed. "All in the family have started a letter-writing campaign to anyone we think might help us find him, or at least find out what happened. But so far there's been nothing. He seems to have just disappeared out in the American West. Mam took it pretty hard. We all did . . ." Tom's voice trailed off.

Annie was touched, but didn't know what to say in response.

"Oh, I am so sorry. Please forgive me for dredging up these emotions concerning my brother. I just got caught up in our conversation. You have that effect on me, Miss Dale." He gave her a teasing smile. "Shame on you for letting me go on like that!"

"I'm so sorry about your brother, Mr. Wright," she said in a most empathetic tone.

"Thank you. I have promised my family that when Lydia and I do go to America, I will find out what happened to George."

✢ · ✢

"Excuse me sir, but do ya know where I might find Tom Wright?" Lydia's twelve-year-old brother Jamesy asked at the Woodhouse Estate gatehouse. "I been told he's workin' here—it's a matter o' life and death that I speak to him."

After a lengthy inquisition, the disgruntled gate guard acquiesced and ordered Tobin to escort Jamesy to the kitchen service entry. Tilly brought him immediately to Tom, who was installing the last of the dumbwaiters for an upcoming dinner party in the main dining hall the following day.

"Mr. Wright," Jamesy, out of breath, approached Tom.

"What are you doing here, lad?" Tom asked. "Is everything all right?"

"I been asked to fetch ya, Mr. Wright. Lydia's real sick!" Jamesy rushed on to get out the words, still trying to catch his breath. "Doc says she ain't got long for this world."

"What? What are ya talkin' about?" The color drained from Tom's face. He immediately set aside his work. "Slow down, Jamesy, and tell me what's goin' on."

"Doc says she's dying, Mr. Wright."

"Ridiculous!" Tom interrupted. "I just saw her not two weeks ago. I didn't even know she was home."

"My sister is dying." Jamesy stood frozen, afraid at seeing the terrifying blaze in Tom's eyes.

Tom didn't wait for further explanation. He packed up his tools and prepared to leave, as Tilly looked on.

"I'm sorry, Tilly, but I have to go."

"You go. Just leave everything, and I'll put it aside for you," Tilly said, nodding toward the door.

Tom's eyes were wild with concern, and he turned to leave without response just as Lord Fitzwilliam stepped in from his chancellery to see what all the ruckus was about. "Mr. Wright, where do you think you're going?" he demanded.

"I'm sorry, Your Lordship," he bowed slightly, panic-stricken as he started for the door.

"You can see the girl tomorrow," he said, then demanded, "You've got to finish up your work here and clean up this mess."

"I gotta go."

"Don't defy me, young man," an indignant Lord Fitzwilliam called after him. "Don't you leave here."

But Tom was gone before the last words left his lips.

Chapter 16

⚓ · ⚓

Tom clambered up the porch steps of Lydia's family shanty. He wiped at the sweat on his face with his grease-stained sleeve. He stood frozen on the barren porch in front of the paintless door to the little hovel. Taking a deep breath to calm his nerves, he prepared to knock, but without warning, the door opened.

"Mr. Wright? I'm Doctor Blackburn," a slim, sallow-faced young man introduced himself.

Tom stood silent. His hands shook, eyes wide with concern. He stared confused at Doctor Blackburn then peered into the darkened room behind him.

The young doctor continued. "I'm sorry, Mr. Wright, but I'm afraid it doesn't look good. It seems phossy poisoning from the match factory has brought on a cancer . . ."

Tom lost the doctor's words after *cancer*. Heat rose to his face. His mouth went dry.

Doctor Blackburn ran on nervously. "I'm afraid . . ."

"Get out of my way!" Tom snapped, glaring at the young doctor. "You don' know what you're talking about." Grabbing the doctor's arm, he pushed him aside.

"She's in God's hands now," Doctor Blackburn tried to console Tom as he was whisked aside.

"Don't! Don't talk to me about God's hands," Tom growled, pushing past him in a surge of fury. "I've had enough of God's will for a lifetime. What kind of God would allow this to—?"

Tom stopped cold, stunned into silence by the sight of the gaunt face of Lydia's mam staring up at him vacantly from the threadbare old couch. She looked much older than her years. There was not the slightest glimmer of hope in her haggard gaze, only defeated sadness. Her desolate eyes set a flat, disillusioned expression of a mother who had given up.

"I'm sorry," Tom whispered.

Lydia's mother didn't reply.

It seemed like Tom could hear the blood rushing in and out of his reluctantly beating heart. His hands shook and he felt woozy.

"She can't be dying?" Tom stammered out. Shame surged through him. "She's . . . she's so full of life. This can't be. Doctors make mistakes, don't they? They're not gods." But he found no hope in the shadowed pall of the musty room.

Doctor Blackburn spoke from the doorway. "I've given her plenty of laudanum to ease the pain, but it's only a matter of time now."

Tom jerked away and turned to the bedroom door. He clenched his fists to stop his hands from shaking. Dredging up what calm he could, he slowly turned the knob and entered Lydia's gloomy room.

When his desperate eyes met hers, he tried hard to control his erratic breathing, his fear, his trembling hands. Everything inside him was screaming, *This isn't fair, it must be a mistake!*

Lydia's beautiful face was gaunt, drained of life's blood. Her eyes were sunk deep into concentric, purple circles. Her cheeks were chalky grey and her lips colorless.

Tom forced a vacant smile. "Hi." He felt like he was underwater, watching the scene unfold from afar. Slowly, deliberately, he took a seat by Lydia's bedside.

She had always been his sword and shield, his comfort and strength. Now she looked so fragile. The beauty of the woman he had asked to marry him on a spring picnic just weeks before was no more.

Despite his pathetic attempt to hide it, Lydia clearly read the panic in Tom's eyes. "Oh, Tom. I'm so very, very sorry," came her apology in a hollow voice.

"This doctor doesn't know everything." He was sure she could hear his pounding heart, like pistons pumping water from the mine.

Lydia smiled wanly.

Hastily, Tom grasped her limp hand lying on the coverlet, frightened anew by its chill. Without the necessity of words, the soft touch of his hand brought a look of concern to her eyes.

"How are you feeling?"

She swallowed hard. "This kinda messes up your plans, doesn't it, Tommy? I'm so sorry. But it doesn't hurt—really, it doesn't. It's like falling in slow motion, only after a while you kinda wish you would hit the ground already."

"Yeah . . ." he nodded in affirmation, still unable to fathom what was happening. He forced himself to hold back the tears, while everything screamed in agony; it seemed his heart would pound right through his chest. Balling his trembling hands, he took another ragged breath to suppress the acid taste of a rising panic.

"You don't know what I'm talking about, do you, Tommy?" she chided, trying to coerce a smile. "You've never fallen off anything in your whole life, you exasperating coal-jockey." With effort, she lifted her hand to his cheek.

"Oh, that's where you're wrong," he answered back, remembering how he felt the day he first met her in the forest.

She smiled with understanding. "We both fell off that cliff, didn't we, Tommy?" She paused to catch her breath.

"I'd rather talk about getting you better, Lydia," he implored.

"Don't you think I tried to talk my way out of this? I'd give anything in the world to change this if I could. I can't bear to see you like this."

"But—"

"No, Tommy. That's not the way we're going to do it. I'm not going to get better," she scolded, soft but firm. "I thought it was a cold or the flu, not the phosphorous inside me. I'm so sorry, Tommy. You deserve better."

Tom sat up in his chair, shaking his head to clear it of this awful verdict. Every muscle in his body tightened, but he couldn't say more without coming apart.

Lydia turned toward the partially curtained window. "Did you know I always wanted to learn to play the flute?" Her listless eyes gazed out the window to the dark clouds passing overhead. "At least, I wanted to learn to play some musical instrument someday. But—"

"Do we really have to talk about falling, flutes, or music right now?"

Lydia flashed with breathless frustration. "What do you want to talk about, Tommy—funerals?" Trying to soften her voice, she rushed on. "Oh, Thomas. I told Mam you'd be okay with an Anglican service. It would mean a lot to her. Is that okay with you?"

Tom's mouth was dry. He drew in a ragged breath. Barely conscious of his words, he forced himself to answer. "Whatever you want."

"What I want is time—time you can't give me," she replied, despondent. "What I wanted, Tommy, was to be your bride. But we have what we have . . . and that's all we have. It hurts too much to think about what I want, Tommy!"

Wiping a tear from her eye, she looked down, then turned back to lock onto his eyes, forcing herself to lift her spirits. "So, entertain me, Pokey. Show me a good time for what I have left of it."

"If only we would have gotten married. Gotten you out of here years ago," he exclaimed in desperation. "Oh, Lydia, I don't think I can—"

"Nonsense, Tommy," she interrupted. "Stop blaming yourself. It's not your fault. I'm going to ask you to promise . . ." She lay back panting and coughing.

"Oh, Lydia, don't! Please, I don't—"

"Now stop it, mister. This is what I worry about when I lie here at night," she said. "I don't want to worry about you. My work with you will be left unfinished, and it's not my fault. You've got to get on with your life . . . or I'll never forgive you." Lydia coughed hard, clearly exhausted. "You gotta promise me you'll get past this," she continued. "Relieve me of my worry. Don't you see? You . . ." She fell back again, weak, exhausted.

"Take it easy, honey." He could do nothing but sit stunned, bewildered, looking at her in concern, his heart pounding. "Please!"

"I look in your gloomy eyes, Tommy, and I still see that determination. You think you can fix this? Well, you can't. You must promise you'll get past this and go on without me. I'm counting on you . . . to release me from this guilt."

Until that moment, Tom had refused to see there was no hope, willing himself to ignore her sunken eyes, the exhaustion and pinched, flaccid complexion he'd seen often in the faces of dying men after a mining accident. The truth surged up from the deepest recesses of his soul—she was indeed dying. *Oh, God, please make my heart stop with hers.*

Tom took another labored breath. "You're right. You deserve better, Lydia. I'll try to pull myself together."

"Please do, Pokey." She stopped to regain her voice. "Promise me . . . you'll get married to some girl. But she can't be as pretty as me!" Lydia tried to smile.

Tom returned a weak one of his own. "It's not possible to find another woman as beautiful as you," he whispered hoarsely.

"You're probably right there." She reached out to touch his hand. "Now, get rid of that sad . . . pathetic face. The doctor doesn't want you depressing me. . . Let's brighten up this place. . . Pull up those shades. . . I don't want to waste what time I have left talking about things we can do nothing about."

"Okay."

She sighed. "I don't care about farms or streams . . . and certainly not pigs. I've been meaning to tell you I never wanted . . . pigs." Lydia limply held Tom by his wrist. "I just want to enjoy the time I have left with you. . . no one else . . . I want to be here with you."

"Of course, you're right. What do you want to talk about?"

"Let's talk about how lucky we've been. . . There are few people who've had what we have . . . in their whole pathetic lifetimes . . . so let's count our many blessings."

Tom was silent for a moment, and then an idea came. "What would you think about us getting married?" Tom spilled out, then got down on one knee. "Will you marry me—tomorrow?"

Lydia smiled with a groan. "I'm not so sure I can be the beautiful bride you were hoping for on your wedding day."

"Honestly, Lydia. There is nothing you could do to change my love for you. I love everything about you. I love that you get cold when you're sitting in front of a raging fire. I love that you call me Pokey, even when everybody knows I'm much faster than you."

Lydia let out a weak chuckle.

"I love the little crinkles you get between your eyes when you're excited. I love that it takes you an hour to get ready while I wait, and you've got only two dresses," he said, wiping the moisture from his eyes. "My love runs as deep as my heartbeat . . . and always will."

"You see there, Tommy? That's what makes it impossible not to love you. Just when I find something annoying about you, you go and say something like this." Lydia wiped the tears of emotion from her eyes. "I so love it when you pander to me. Yes, of course I will marry you, Tommy. I'll marry you any way you want. If you want to drag me out of bed wearing this ugly rag, and sit me in front of some big-eared, pointy-faced minister . . . I'll do it."

"Don't worry, we'll get Bishop Walker. And I'm gonna have Mam take in her wedding dress for you. Edie can help with your hair. Maybe we can even have your little brother find something to rouge your lips?"

Lydia rolled her eyes. "I don't suppose Jamesy could make me look any worse."

"You'll be ravishing!" He reached out and touched her hand sweetly. "We can have the wedding right here with flowers, candles, the works."

Lydia nodded. "I think I'd like that, Tommy. It would give you something to do besides mopin' around . . . and annoying me to death." She reached out and grabbed his hand. "It all sounds divinely wonderful." Lydia wiped away more tears from her eyes. "You can be just so darn sweet sometimes. I'm sure gonna miss you."

Tom sat quiet with Martha as she shortened the hem and tightened the bodice on her wedding dress to fit Lydia's thin, frail frame. As always with her children, Martha would give it her very best. It would take her most of the night to finish, but in the end, the dress would be absolutely beautiful.

"How are you doing, Tom?" she asked as she stitched.

Tom was quiet for a long moment, and then decided to be honest with her. "I could tell you I'm fine, but that would be a lie."

"It will be difficult for you. But though you may feel you have a heart broken beyond repair, never forget that even a broken heart in the hands of a Divine Healer can be made better. Like a broken bone in our miraculous body, it will heal if you let it."

Tom swallowed hard. "I'm not so sure I believe that right now, Mam. How could a loving God allow this to happen? I've lost everything. And if He is up there looking down on me, I hate Him for it."

Martha took on a sober look. "You blame God for what happened to Lydia?"

"Yeah, I do." Then, thinking of his mam's religious faith, Tom qualified his answer with, "Maybe?"

She frowned.

Tom shook his head. "I'm sorry. I don't mean to offend you."

"Why is it that we frequently blame God for what goes wrong in our lives, but don't give Him credit for what goes right?" She gave Tom a moment to digest the question. "Your love of Lydia? A once-in-a-lifetime experience. Do you think God had anything to do with that?"

Tom looked down at his feet, pondering her words, feeling ashamed. "I don't know."

"I'm guessing you wouldn't give up those wonderful years you had with Lydia for anything in the world. She enriched your life and put you on a path to becoming a better man than you were." She paused, letting him digest her words. "It seems to me that the only way to truly take your sorrow out of Lydia's death would be to take your love for her out of your life."

Martha, always insightful, seemed to know what to say and what to do, but this time was difficult. He'd had this very conversation with Lydia. Her death would deeply affect the entire family. Both families were reconciled now. It would not be long before Lydia was gone from their lives, but not for Tom. Never for him.

"I'm sorry, Mam, but that doesn't ease the pain. And despite Lydia's request, I can't ever imagine myself moving on from this."

Martha pulled the dress bodice into the lantern light to inspect her sewing. "Tell me what you will miss most of all."

"Well, I suppose there are a million things I will miss." He paused reflectively. "Mostly, it will be the small and simple things. It felt like we were destined for each other the first time I met her on that path in the woods. The first time I touched her hand. And every time we walked to our private meadow on Sunday afternoons, I knew we were meant to be together, like I knew my own name."

Martha's eyes moistened. "What will you do now?"

"I really don't know, Mam. I suppose I'll get up in the morning and blindly walk through the day. Then do the same the next day until, hopefully, I don't have to remind myself to get out of bed." He paused. "But I don't really know."

Recognizing the tremendous struggle that he was about to face, Martha added, "Your grief will never recall her back to you. But as with all of us in our long process of becoming, the way you live your life after she's gone might revive the best of her memory to the world. Don't allow yourself to be a prisoner in your own mind."

"I don't know if I can do that," Tom answered, his voice trembling.

"A little bit of her soul will always live along with you, Thomas. Don't let her down. Don't lose hope."

"It's so hard, Mam."

"You must keep hope alive. We all must. Hope is essential if any of us are ever to survive the heartache life throws in our path. You must either get on with your life or just give up on it. Lydia would pray you chose the former."

➤ • ◄

Thomas Wright and Lydia Kaye were married on a beautiful spring day in May by Bishop Walker. Lydia lay in her bed tenuously holding onto life. She was dressed in the newly fitted wedding gown, courtesy of Martha. They were surrounded by candles, those they loved, and wildflowers from their special place in the meadow.

What little time they had left on this earth would be spent together. In the quiet of the night, they held each other tight, refusing to be separated all the night long.

In the hours before dawn, Lydia lay watching Tom fidget in his sleep beside her. He looked so exhausted. Oh, how she loved him. She was concerned as much about what would become of him after she had gone as she was disappointed at leaving him to fend for himself.

He opened his eyes to his new wife and smiled away the sleep.

"I'm gonna miss you, Pokey," she whispered. "Ya know, you men are always hopeful we women will forget . . . and we women are always hopeful you will remember."

"What is it that you are concerned I won't remember, my wife?"

"You have to promise me you'll get on with your life, Tommy." Patiently, lovingly, she looked deeply into his eyes. "Remarry and go to America. Promise me you'll have a family to love and bring meaning to your life," she pleaded. "Home is where the heart is, and happiness is not far away." She put her hand over his. "Promise me."

"I promise," he said reluctantly, if only to ease her worry at the last.

"Remember that promise. . . I'm gonna keep my eye on you, Tommy Wright . . . from wherever it is I end up. . . So, don't try to shirk your promise, 'cause I will haunt you if you do."

Tom gazed back at her with a halfhearted smile, but could say nothing more.

Chapter 17

✈ · ✈

THREE DAYS AFTER THEIR WEDDING, Tom sat on Lydia's straw
mattress, her head in his lap. He had watched his wife as she
slept all that long night, her breathing shallow. It had been raspy for
about an hour. He knew he would never again hold her in this life.

For a moment she opened her eyes. She smiled weakly up at him
with the reflection of moonlight on her face.

He stroked her head and whispered, "You know I was warned not
to marry such a beautiful woman."

"Warned by who?"

"Everyone with eyesight!"

Lydia sighed. "I'm gonna miss many things, Tommy," she whispered in a ragged voice.

"The sunlight rising over the hills at daybreak to coax open
springtime blossoms. The smell of coffee brewing in Tilly's kitchen
on a cold morning. Pulling the covers up under my chin after a long,
hard day of work."

"We don't always appreciate the simple beauty of this world,
do we?"

"Until we're about to lose it." She went on, trying to get it all out. "But most of all, I'm gonna miss you. The dreams for our future. You holding me like this when I'm scared. I'm only sorry you never made love to me." She paused, then went on, faintly. "I'm afraid of what the morrow will bring . . . I do love you so . . . Thank you. Thank you for all of it." The edges of Lydia's voice fluttered to silence.

Exhausted, she faintly smiled one last time. Then she closed her eyes, quietly falling asleep in his arms. She inhaled deeply, holding it in for three or four long seconds, then exhaled in a long, drawn-out sigh of relief.

Several seconds passed before she inhaled again. Again, she held it. Again, the deep sigh, as if something down deep inside her was being released.

Thomas found himself holding his own labored breath, willing her to take yet another.

Then he felt her body, almost imperceptibly, relax. She settled more deeply into his arms. There was one more breath. One more last long sigh. And then it was over.

He held her until death loosened her grasp. His Lydia was gone.

For a long time, Tom sat with her in the quiet of the night, alone with the woman he would forever love. Tears streamed down his cheeks.

"I suppose this is the end of a road neither of us even considered traveling, my love." He bent down and kissed her forehead, wet from his tears, still, and growing cold.

His beautiful young wife died before her life really began. He could not forgive those responsible for taking the one he loved most of all. Martha's voice rang in his head, "'These arrogant lords and ladies cast aside all conscience and morality to squeeze out every last shilling from our lives for their comfort, their pleasure, their extravagance. Damn them all to hell!'"

⊱ • ⊰

Frequently death comes as an intruder. It is no respecter of persons. Its summons is often heard before scarcely reaching the beginning in life's journey, cutting short the joy of expectation, shutting off the light to a promising future. It invades as an enemy in the midst of life's feast, an unwelcomed vanquisher of human happiness.

Tom gave all he'd saved for their immigration to Lydia's family for her funeral. But he could not bring himself to look at caskets, tombstones, or meet with pastors to discuss cemetery plots.

He stood alone at the funeral before her open pinewood casket, head bowed in solemn reverence. Death had laid its heavy hand on Tom's shoulders. It left all around him baffled, bewildered, wondering what could be done to ease the pain of this once strong, vibrant, and driven young man with all the hopes and dreams for the bounties of life. Now, all those dreams had been crushed and mauled, leaving him a broken shell of the man he once was.

He looked at what remained of her mortal body. It was her whom he loved so deeply . . . and yet it was not her. There was something missing. Her countenance had lost that special something that made her who she was, that vibrance of life. He stood there frozen, hardly able to breathe.

Something in her innermost soul was gone. The look when she smiled at him. The mischievous sparkle in her eyes, her uniquely vivacious, contagious laughter when she teased him, the dance in her eyes when she spoke excitedly.

Tom placed their marriage certificate, issued just the week before, and his wedding ring signifying his circle of promise inside the desolate pinewood casket containing her lifeless body. It was all he had to prove that he had once been hers and she his. He caressed her face one last time. But when he touched her skin, he hesitated. Her warmth, her love of life, was gone. She was cold, distant, no longer there. The body lying in the coffin was not Lydia. He felt guilty for dishonoring her. Oh, how he wished he were with her now in some heavenly place. Then his own insulting words echoed painfully in his ears: *"Who needs an afterlife? I'm strong enough to achieve my own goals in this life without making weak excuses for failure. Heaven is just a place for the feeble-minded to deposit their troubles."*

He looked away, appalled at his self-centered arrogance. Ashamed, he stepped back, and the minister began his eulogy.

"For as much as it pleases God in His great mercy to take unto Himself the soul of this dear sister," said the minister, "we commend to the Almighty our beloved Lydia Kaye Wright. We commit her body to the ground. Earth to earth; ashes to ashes; dust to dust. May the Lord bless her and keep her. May the Lord make His face to shine upon her and be gracious unto her and give her peace forever. Amen."

The carpenter nailed down the lid to the coffin, one nail at a time, until the slamming down of the final nail closed in Tom's soulmate until the end of time. Tom watched as the casket was lowered into that awful pit. He felt the haunting devastation wrench his heart as the earth was shoveled in to bury her forever. She would never again be by his side—a recognition that seemed to tear his heart from his chest.

"For as much as it pleases God," the minister had said.

How could this please God? Tom cried out in his heart.

Chapter 18

⋟ · ⋞

TOM SAT ON THE EDGE of his rickety bed, alone in his ten-by-ten room cloaked in silence. He had told Martha after the funeral he wanted to be by himself, to wallow in his sorrows, away from everything and everyone. He didn't want to see family or friends, share in their condolences, or hear their platitudes designed to lift his spirits.

"Please forgive me," he'd requested of Martha, "but I don't want to have to bite my tongue to avoid lashing out. Please, no more tears of sorrow for me. Just let me be depressed, despondent, and sullen for a time."

All Tom wanted was a quiet place to embrace the pain of his loss, to be a solitary recluse. He'd been drinking brandy all afternoon and into the evening to smooth out the harsh contours of remorse from the long day that seemed to drag on endlessly.

Feeling exhausted, Tom's thoughts ticked away mechanically to the lonely clock of his mind. There was a blunt dullness pushing back on his misery, against the rising tide of pain now to come. Like severed tissues, shocked by the surgeon's knife, there had been a brief instant with comparatively manageable anguish before the real grief came at him hard. When the dullness wore off, it gave way to the sharp pain of final recognition that he would never again see her, never again hear her voice, never again feel her touch.

"No . . . no, Lydia, no!" he thundered into the night. "Why did you leave me like this? You were supposed to be different. You were supposed to never leave me alone again. But I couldn't even trust you."

Tom put his head in his hands, drugged by despair, questioning what he should have done differently, wondering why. His most poignant memories would be forever tainted by some arrogant aristocrat who had taken from him those whom he loved most—his grand, Tigger, and now his precious Lydia.

A loud knock at the door jolted him back into the moment. He tried to ignore it. He was in no mood for socializing.

"Go away," he shouted.

The knock came again, and then louder the third time. Irritated, he stood, moving in a daze toward the door. But before he could open it, the pounding began again in earnest. He yanked open the door.

"I'm sorry for intrudin' on your grief, Mr. Wright," Mick Hemmings, a tall, skinny, contrite drink of water stuttered out, as he glanced back. "But Mr. Black here insisted I—"

"Whatta you want?" Tom interrupted.

"Enough of this jawin'."

Candyman Boo Black pushed passed Hemmings and clomped into the room in his wet, muddy boots, his fat belly hanging over his belt. He grunted, looking around the room, puffing on his cigar. Then he coldly turned to Tom and sneered, "I hears ya walked out on His Lordship without so much as a 'by your leave.'" With a one-eyed stare, Boo blew cigar smoke in Tom's direction through his nicotine-stained lips. "What we got here, Wright, is a failure to give due respect to your betters, and by God, I won't have it."

Boo Black's sardonic grin proudly displayed his two silver front teeth in a graveyard full of sparsely spaced, putrefied rotting ones.

"Ya been puttin' on airs far too long. Fraternizin' above your station, fancyin' yerself the miners' go-to man," he snarled. "Thinkin' you're somebody special with all that book learnin'."

Fury rose within Tom, his heart pounding, his hands closed into tight fists, his face reddened in cold, rising anger. There had been a long line of unjustifiable abuse from Boo Black. He'd unleashed his venomous brutality on too many miners.

"Ain't no room 'round here for your influencin' the men into thinkin' they're deservin' somethin' better than they's gettin'. I'm gonna enjoy stickin' it to ya." Boo's deranged smile of satisfaction spread wide across his face. Then he forced his brows together over his single, maniacal eye above a pronounced frown. "I'll bring the wrath of God down on your sorry ass." He snarled, slamming his fist so hard against the wall that Mick Hemmings jumped, and cowed in submission.

Without warning, Tom turned his attention from Mick to look directly into the sadistic eye of the Candyman. His icy cold stare caused Boo to step back. Defensively, Boo tried to regain his composure, putting his best face of condescension forward. He strode arrogantly across the freshly scrubbed floors in his muddy boots, puffing on his obnoxious cigar. Then he turned in a vulgar glare toward Tom and blew out a long puff of smoke. It was clear Boo Black was waiting for verbal acknowledgement of his reprimand.

Mick took another step backward, into the corner, to get out of the way of what seemed an inevitable confrontation between these two glaring men.

Throwing the remnants of his cigar on the floor, Boo ground his muddy boot on the butt. "Well, there ain't nothin' ya can do 'bout it now, Wright. She's gone. I'll be expectin' to see ya first thing in the morning, bright-eyed and bushy-tailed! Ya'll be docked for time missed, o'course." Boo Black spat a slug of tobacco toward the garbage can, but it missed, splattering across the wooden floor. "Of course, we'll be expectin' an apology to His Lordship. And iffen ya don't show, well, then don't be expectin' to have a job when ya come a crawlin' back."

The demand was a miscalculated mistake, like carelessly lighting a match near a keg of black powder.

Tom Wright had worked in the mine since he was a boy, but he had grown to be a mature, seasoned, confident man. Blessed at birth with the slow-to-anger nature of his father, he had dealt with life without direct confrontation. He had simply endured the cold, miserable conditions, the backbreaking work, the poor pay while doing what he could to assuage the vast injustices of the system, but always careful to avoid provoking the monstrous evil directly. And in recent years he had only been biding his time until he could afford to leave his family safely behind, and the oppressive system altogether.

But unfortunately for Boo Black, Tom also had a streak of his mother in him. And on this night, something snapped in Tom.

Tall now, with a hard-muscled back and shoulders from his early years working in the mine, Tom moved to stand directly in front of Boo. He looked down on the pathetic man with fire in his eyes.

Boo took a step back. Then another.

It wasn't enough. Tom stepped forward to fill the space between them. He leaned in, forcing Boo's back up against the wall. He grabbed Boo Black by the lapels, lifted him off the floor, and slammed him against the wall with a *thud* so that they were eye to eye.

Boo's face went ghastly grey. He was sweating profusely, his eye wide. Unable to speak, he hung in the air, pinned to the wall like the blood-drained carcass of a deer ready for slaughter.

When Tom spoke, his voice was calm, low, and menacing. With his nose nearly touching Boo's fat, chalky-grey face, he slowly made his point, "I won't be coming to work in the morning. Am I clear enough for you, Mr. Black?"

Boo humbly nodded in affirmation.

"I didn't hear you, Mr. Black."

"That'll be fine!" Boo confirmed. Both Tom and Mick could see the fear in his eyes.

"Good!" Tom dropped the despicable man to the floor. "Now get the hell outta my house!"

Boo, his head down in submission, hurried into the night.

"I don't ever want to see you around here again," Tom called out as Boo, his eyes on his feet, scurried down the path without another word.

Mick stood stunned in the far corner of the room with unabashed appreciation—no, more like adulation. He had been witness to what every miner had fanaticized.

Meekly, he started down the steps, but turned back with his hand on the doorknob to look toward Tom.

"That . . . well, that was simply amazin', Mr. Wright." Mick looked on him admiringly. "How did it feel?"

Tom did not respond, but continued his cold, hard stare toward the humiliated figure of Boo Black slinking into the night.

"All right then, Mr. Wright, have a nice rest of your night. And may I thank ya for this fine evenin'!" Mick tipped his hat, and with that, he too was gone.

Tom knew Boo Black would do everything within his power to avenge this humiliation, witnessed by one of the miners he was charged to keep in line. Neither the company, nor Lord Fitzwilliam, could afford to allow Tom to humiliate their enforcer—no matter how incompetent Boo was. He was certain the price for his insolence would be high and likely fatal.

The thought skittered down

➤ • ◀

In the bleak hours long after Boo Black left, Tom quietly sat leaning over his desk amid the debris of his rage. Still as a marble bust, Tom's jaw was clenched tight, his brows knit together in concentration, his desolate eyes focused downward, framed by a sweaty tangle of his wavy hair. Grand's flintlock pistol lay on the desk before him.

He was almost afraid to breathe after the war-torn night of anger and despair, still awash in the unbearable pain of knowing he would never again be with Lydia, never again hear her voice, never again feel her touch. He was not about to give Boo Black or the company the satisfaction of taking his life in retribution for his unpardonable sin of loving his wife beyond life itself.

The easy way out of his misery sat right before him. Why shouldn't he end it all here?

In the shadowy darkness of the flickering candlelit night, he passed his hand over the shiny barrel of the pistol, then brushed its smooth wooden stock. His hand settled on the triggered hammer notched atop the lock-plate of cold, hard steel. He was bewitched. The notches seemed to point at him like the knuckles of an accusing finger. In the flickering candlelight, Tom glared at the pistol—a terrible, wicked thing. A deadly, avenging thing. *It would be so simple to join her*, he thought.

A chill ran down his arms to his fingertips and he jerked from his evil trance in a cold sweat, desperately gulping for air, shuddering. The rush of adrenaline set his heart to beating wildly. He was frantic, full of dread. He let out a rattled breath and leaned back in his chair. Slowly, the torment raging through his veins began to subside. Staring at the pistol, he wondered . . . *just how many footsteps are we away from our own demise, really?* The thought skittered down his spine, brushing away the last tendrils of insanity. He reached down, pushing his hand past the pistol, and picking up the compass Grand had also left him. He would not take the coward's way out of his nightmare.

The spell broken, he pulled away from the treacherous abyss, forcing a calm in his spirit and dredging up the strength to go on.

Chapter 19

FOR THE FIRST TIME IN his life, Tom would not face up to his problems. He would abandon all prudence and reason and leave his troubles behind—his family, his friends, the village, and the pending death warrant from the company enforcer.

Tom went through the cupboards, drawers, and closet, stuffing what food, gear, and other supplies he could carry into his backpack. He gazed longingly at his many precious books, windows to the outside world. He would miss them, but he knew they would have to be left behind with all else. He strapped his bow and quiver of arrows across his chest, closed the door to his cottage behind him, and headed down the still-dark lane and out of the village.

Tom had no real destination in mind, no idea how far he might go, or what he might do when he got there. He knew if he stayed, neither he nor his family would be safe. He needed to go far away to a place where he could be alone. He was done with being responsible, dependable, or steady under fire. With a breaking heart, he took one step after another in a mind-numbing rhythm mile after mile, with a vague hope of outrunning his tormented heart and his all-encompassing sadness.

He reached the outskirts of Elsecar long before the sun rose in the east. He continued through Shire Brook meadows, the ancient village of Sharrow Edge, and on to the quaint twelfth-century town of Crosspool.

By nightfall, Tom was far from home, heading toward the centuries-old hamlets of Bamford, Castleton, and Heathersage where Robin Hood once roamed. Turning north, he headed up into Hope Valley. The night rolled into another day, one bewildering step blending into the next. Far from the last vestiges of humanity, he wandered deeper into ancient river valleys of woodland forest. He dared not sleep, for fear of his demons creeping into his mind like thieves in the night.

By the morning of the second day, he had reached the high mountain meadows surrounded by the staggering Yorkshire peaks, untouched by man for centuries. In late morning the first summer storm slipped over the mountains. With it came a stubborn fog rolling into the hollows, twisting its tendrils over and around the hillsides. It crept through the woodland and settled in the meadow where the grasses, cloaked in deep greens, bowed their heads with the rain. All through the afternoon, the storm gathered strength, runoff swelling the normally peaceful brooks to their banks. The water rushed impatiently along a contoured course through the meadow toward the hollows below. Still, Tom ventured ever deeper into the forest, following an eagerly flowing brook. The pounding rain grew harder. All the animals had nestled snug in their homes to weather the storm.

For most of that day, Tom had been on the lookout for someplace to shelter. He was wet and cold—the kind of biting cold that temperatures don't accurately reflect. He shivered uncontrollably. His back pained him. His legs ached. His feet were numb. And though his stomach growled in complaint, he was too tired to do anything about his hunger.

Then he saw it—a glimmer of light passing through the clouds at dusk, reflected off a modeled scarp in the jagged limestone cliffs perched high above an alluvial fan that spread out from the base of the hills. The notch in the light-colored limestone was barely visible in the heavily vegetated slope.

Tom slipped, slid, scrabbled, and climbed his way up the slope in driving rain pearling from clouds, to mountain tops and down the slopes. Streaks of lightning raced across the sky, chased by cracks of thunder. He reached the plateau and found the cleft in the rock was the camouflaged entrance into an ancient limestone mine, covered with brambles of bilberry, rock rose, dark red helleborine, and privet vine grown up over the centuries.

With his hand ax, Tom hacked away at the heavy foliage until he could see the discolored limestone walls within. Soaked, rainwater running down his face, he cleared an opening and slid in out of the rain. A shiver ran down his spine as he swept aside the feathered roots that hung from the cavern ceiling in the shadowed darkness.

"Hello! Is anyone in here?" he called into the cave. Cautiously, he looked over his shoulder, as though expecting to find some predatory animal watching from the shadows.

He passed his hand over the stone clefts in the limestone walls. "Quarried," he muttered. Centuries ago, miners, not so unlike himself, must have spiked and cut limestone block in a once-bustling quarry to build the magnificent castles in Heathersage or the Castleton valley below. Perhaps it was limestone for the ancient Roman foundations on which those castles were now built. He couldn't help but imagine those long-ago craftsmen, their hardened copper chisels and iron hammers working these cavern walls.

"In some ways," Tom whispered, "things haven't changed all that much. We still work as serfs for a noble class." But the centuries had reclaimed this stone quarry, returning it again to nature's hold.

Tom wiped the sweat from his brow and slumped down to the stone floor, exhausted. He leaned back against the limestone wall, spreading his legs wide apart and resting his hands on the floor matted with debris. Drained of all remaining strength, he felt the loneliness push on him. His heart was heavy with the hopelessness of someone who had nowhere else to turn.

"What do I do now?" With agony of heart, he raised his eyes to the heavens and pleaded, "Oh, God. I'm not a praying man. But if You're up there, I sure could use Your help." Tears streamed down his cheeks. "No one stays. Not even Lydia. Please, I beg of You, stay this night with me."

Slumping back into the corner, he closed his eyes where he sat. Dusk fell into night. The chattering winds whipped through the hollows, rattling the trees. The chorus of the storm and the beat of falling rain just outside the protection of his shelter enticed his tired body to give up the struggle. The rhythm of his pulse seemed in sync with the cadence of his breath—coming in and going out, slow and even. His fatigued muscles twitched and shuddered as the curtain of night fell in around him, and he slipped into a restless slumber for the first time in days.

Now the demons of his fitful mind could no longer be kept at bay. Dark thoughts pulsed inside him, as though they had waited for his defenses to be lowered. Like some great beast they breathed hard at the windowpanes of his mind. The crying buzzards of his greatest fears circled about in a whirlwind, swooping down with outstretched talons and sharp beaks, tearing into the flesh of his nightmare, ripping at the heart of where his fears lived.

In the darkest corridors of his mind Lydia's face appeared from inside her coffin box, suffering from nature's course, rising to confront him. The worms of the earth had been at work, robbing her of her beauty. Tom tried to pull his subconscious eyes away from the awful vision. In a cold sweat, frantically gulping for air, he twisted, turned, and writhed, away from Lydia's disfigured features, trying to return her to the earth, ashes to ashes, dust to dust. But as he pulled back from his awful dreams, he saw a vision of himself lying beside her—decaying, decomposing, and festering just as she was. Dark, suffocating shadows pressed down upon his chest. And in his nightmare, he recognized all may not be just the imagination of a sleep-deprived mind, but rather a premonition of things to come, his own demise.

The gloomy despair Tom suffered of mind and spirit was more than the mere heartache of loss. It was a place of hopelessness and depression where rational thought no longer held sway on the human spirit. The vast loneliness grew roots inside him, encircling his heart in a suffocating grip.

The winds blew. Lightning ripped open the sky, unleashing a downpour of rain into the hollows. And Tom slept fitfully all through the stormy night, not knowing what the morrow would bring.

Chapter 20

★ · ★

CAPTIVATED BY THE SLOW GLIDE of a peregrine falcon over the meadow and brook running through it, Tom sat silent at the entrance to his cavern home of the past six months. He warmed his hands on a hot cup of evergreen tea. Steam rose in the cold, peaceful morning air as the first light of dawn crept into Spring Hollow, sheltered in the protective cradle of the imposing Peak Mountains.

Over the summer and fall he had thrown himself into the work of survival. In this quiet asylum, far from the complications of humanity, he put aside the calamities of the outside world. A lone man in the wilderness, he had quarantined himself, and with blood, sweat, tears, and surety of wit, he had survived and even prospered. There were no eyes upon him here in the harsh environs of nature, except those of the Almighty. And for that he'd been grateful.

It seemed on this calm, cold morning all had changed overnight. The predawn whistling winds had swept through Spring Hollow, clearing away the last vestiges of autumn. The evergreens may not have noticed, but the sycamore surely did. Perched precariously at water's edge off the pooling pond, the giant sycamore's yellow, red, and golden autumn leaves had broken fee to fly in a blaze of color, swirling and sailing in the cold wind on their last dance to beckon the start of winter.

His eyes followed the current of water meandering in the brook on its tranquil course through the meadow to pond in the life-giving estuary. The pooling pond, created by hardworking beavers in building their logjam home, was the cornerstone in the circle life of the meadow. In nature's own way, the destructive power of the raging water had been tamed to preserve and support the abundance of life.

Thanks to the beaver, whom Tom affectionately named Sylvester, Tom also had all the logs he had needed to build onto his new cavern home, complete with a protective portico and insulated pine-bough door to keep out the cold night air.

Inside the temperature remained relatively constant. He had fashioned a mat of dried reeds to cover the rough-hewn limestone floor, a deerskin mattress stuffed with pine boughs, and a heavy rabbit-fur blanket. A haunch of venison hung now from the ceiling, and in the corner sat a wooden box filled with honeycomb to make honeyberry jam. Lining the walls were tools, animal skins, and a formidable array of hunting traps, snares, a bow, an obsidian-tipped spear, and knives. The rear of the comfortable grotto home was partitioned off with cords of drying firewood, and behind that lay the cave. The black void led into a catacomb of ancient tunnels, some of which he still had not explored.

A cooking hearth had been built of flat stones in another corner of the den; a carved bowl with utensils sat beside it. On cold mornings, when the hearth glowed with fire, the surrounding stacked rock provided thermal warmth that radiated throughout the room. Tendrils of smoke were drawn up from the hearth through a chimney fashioned of stone and timber, mortared together with crushed limestone cement. A warming grill stationed high above the fire held freshly smoked trout. A cooking pot, filled with a simmering breakfast of rabbit stew, hung on an arm from the limestone wall, and the pungent smell of herbs from the glen filled the air.

For weeks after Tom's arrival, the miracle of fire had remained an elusive mystery. A simple spark and ember, taken for granted in his former life, became the goal for which he spent dozens of hours in pursuit. Using a hickory spindle, he had rubbed so long and hard, it had worn a groove in the cottonwood fire board and left blisters on his hands before he learned the art of creating a spark to ignite an ember, then a kindling fire. At long last, through trial and error, Tom had solved the mystery of the magic, and the secret to drawing smoke up the flue with an ingenious design that brought fresh air into his cavern home. The exhilarating discovery proved to be the pinnacle event in his survival and transformed his daily life with tasty, cooked meals and a fire in the forge to make tools. His diet of crawdads, raw fish, and wheatgrass from the creek was over, replaced by grilled trout, rotisserie rabbit, roasted boar, wild onions, and fresh mushrooms and herbs from the meadow, all cooked to perfection. "A cuisine reserved for the gods," he'd told himself.

The tea warmed him as he drank it down on this cold morning. Staring at the great sycamore perched on the edge of the pooling pond, he marveled at the magnificent canopy of recumbent limbs stretching out over the glassy surface. Like giant knuckles, the ancient roots jutted out into the water well beyond the shoreline. They created little grottos for Sylvester's beaver kits to play and the ducks to feed.

By imperceptibly small increments of growth over decades of whistling winds and drenching rains, this great tree had sent its roots deep into the ground, bolstering its weaknesses, discouraging rot, and strengthening foundational support for its heavy limbs to withstand the forces of nature. Both cold winter storms and warm summer sunshine had driven its roots deeper into the safety of the embankment to nourish, and help hold onto abutment footings and to life itself. Adversity had given this great tree its strength to overcome the challenges it would face during a lifetime of wind, rain, and storm to grow into one of the most magnificent specimens of God's creation.

"I suppose it's not so dissimilar for man?" Tom muttered under his breath.

Each task that required Tom to solve a problem for his daily survival drew his mind away from thoughts of Lydia, suffocating loneliness, the concerns of the outside world, and helped fend off the shroud of despair from taking charge. The work had been hard, taxing on his body, but surprisingly rewarding, strengthening, and thought-provoking. In the clean mountain air of Spring Hollow, his labor was invigorating and it brought a sense of accomplishment and self-reliant gratification. The patient creative ingenuity birthed a peace and confidence that seemed to strengthen him, reshaping his view of the world. It gave him something of a spiritual outlook on nature and his perception of life.

With the work and struggle to survive in nature's sanctuary, Tom had gained a greater appreciation for the gift of life. Like the great sycamore, he had grown roots into a foundation that made him stronger and better able to withstand the whirlwinds, hail, and mighty storms of life intent upon dragging him down.

He was reminded of his conversation with Annie at the butcher's shop. How had Annie put it? "Sometimes, it's by small steps the Almighty does His best work in the icy landscape of winter, during storms of thunder, under punishing winds of trial. Either we rise to the challenge during the hard times, or we collapse under the weight of our burdens."

How insightful she had been when she let her mind run free. Those words seemed particularly profound now, here in this place.

⤝ · ⤚

Dusk settled into darkness on a winter's eve. The thunder roared and the cold north wind howled as it swept across the desolate, snow-covered meadow and whistled through the frozen woodland. The cavernous darkness seemed to be a right-now thing, vast and foreboding. But on this night, the glow from the dancing flames of a fire warmed him and forced the dispiriting blackness to recede into the corners of his consoling home. Wrapped in his warmest fur blanket, Tom poked at the embers to encourage the flames, mustering the resolve to confront his macabre loneliness. He knew the time had come for him to decide whether to continue bearing his poverty of spirit, or to recognize the futility of dwelling on bitter heartache, to push back on debilitating thoughts and get on with the business of living.

The wind calmed, and the snow stopped falling. The crackle and sputter of the fire broke the deafening silence. Tom pushed open the door to look on the world outside. Shadows from flickering flames danced across the snow-covered meadow.

With his knees tucked up under his chin, arms folded over his legs, he wrapped himself tighter into his fur blanket and returned his gaze to the scarlet embers of the fire. He remembered something Annie once said. "Firelight—where all time stops, and we reminisce of times past."

He closed his eyes in the shadows of the den that smelled of living and tried to imagine casting a line back across time, across space to another place, where his family celebrated the humble birth of a babe in a manger at Christmastime. "Christ had come not just for the great men of the ages," Mam had told them as they curled up in front of the fire in the hearth, "but for all men and women, requesting each to love thy neighbor. And that's why we go about serving one another at this time of the year."

Tom smiled, recalling a long-ago Christmas Eve. It was just after Tigger had died in that awful tragedy, along with Tom's two cousins, Abraham and Isaac. He and Mam had stopped by Aunt Mary's with a bowl of porridge on that cold evening. The rest of the family had stayed home to complete the final preparations for a most bounteous Christmas dinner. He and Mam had found Aunt Mary in a straw bed on the floor of their cold, dark shanty. Aunt Mary was huddled under blankets with the baby and her two little girls. They were trying to keep warm in a house with no coal for the fire, or food for their starving bellies. Tommy and Mam returned home still holding the bowl of porridge. Joey and Edie, having just finished setting the table, sat with Papa and little Georgie, arms folded in front of their plates awaiting Mam's return. Their stomachs were growling.

As Mam stood in the doorway and told of Aunt Mary's plight, all in the family listened intently without a word. It was Edie who was the first to speak, "Well then . . . we must give them our Christmas dinner."

Mam, seeing confirming nods from the rest of the family, smiled. "Thank you, my pretties." Then turning to each she asked, "Shall we have the porridge then?"

Warm, tender thoughts of home flooded into Tom's heart. He hadn't cared that his family was poor, that there would be only porridge for Christmas dinner. They had each other. These gifts of love were enough. It was the pure joy of sharing Christmas together as a family in front of a fire, in the humblest of circumstances, that made Christmas special. A smile of remembrance eased across his face. Simply being together as a family had brought them happiness.

He wondered what his family was doing right now. He wondered what he was doing, alone in this grotto, without family, those he loved most. He felt selfish for having not recognized his many blessings. He wished there was no evil in the world—no Boo Black or aristocrats intent upon endangering him and his family. But . . .

"This could be Christmas Eve, Mrs. Shrew," he muttered to his roommate while she busied herself eating bugs; staying close enough to the fire to partake of its warmth. The tiny shrew had moved in with the first snows. He considered himself fortunate to have her, for she bartered her rent by providing a much-needed cleaning service. Nearly blind, she was like a street sweeper, using only her highly sensitive whiskers to locate and devour her prey. She swept up larvae, insect eggs, sheltering crickets, cockroaches, spiders, and even mice as big as herself, eating her own weight almost every day.

"And why shouldn't it be Christmas, Mrs. Shrew?" He paused for a moment to ponder the question. "Frankly, I need it to be Christmas!" And thus he willed it so.

Under the influence of the hypnotic dance and crackle of the flames in the hearth, he struggled mightily to rekindle the bygone Christmas spirit of his youth.

"Mrs. Shrew, please accept my Christmas gift." Tom shared a bit of his rabbit stew supper with her. Then he sat back and watched.

"Manners, Mrs. Shrew. Manners, please," he admonished as she eagerly gulped down her Christmas dinner.

Up on her hind legs, she looked at him, begging for more.

"You know you're a glutton," he said, handing her a bit more. "But you're welcome."

He watched the dying flames flicker and dance over red-hot coals. It had once seemed the only light in these dark times, but he found on this night there was a new flickering that had begun to flood his heart. His pondering had rekindled thoughts of going to

America. Living in the raw wilderness of California's Sierra Nevada mountains. Maybe even hacking out a place in the wilderness for his own family.

Deep in thought, he turned to look out onto the cold, quiet night outside. The clouds had parted. The shimmering glow of moonlight flung hope across the meadow and reflected off the thin, crisp cover of the freshly fallen snow in a kaleidoscope of color. There was neither footprint nor sign of life in the resilient scene. The solemn, sacred beauty of the pristine, snow-covered wilderness was breathtaking, a lonely grandeur to be shared with his eyes only as he focused on happier times and a new future.

⊱ • ⊰

Little by little, the sun was no longer willing to stand by, submissive to the bitter cold and darkness of winter. It had begun to exert its independence more frequently of late, until finally winter relinquished its grip on the meadow and spring elbowed its way in. The days grew warmer, the cool nights shorter, and the wind and rain receded to make room for another season of renewal.

On a crisp spring morning, Thomas stood outside his cavern home, still and quiet. He gazed out at the scattered oaks on gentle hillsides, twisted willows lining the meandering brook, and the giant sycamore spreading its recumbent limbs over the pooling pond. The soft light of daybreak crept unnoticed over the mountains, bringing in the warmth of a new day to hillsides filled with the moisture of spring thaw. The atoning sun rose luminescent on the springtime of life, sending its saving grace through the pillowy grey clouds that drifted easterly on distant sea breezes. It felt good to begin life anew with a plan.

He stretched his arms out wide, and high overhead. He arched his back, breathing in the crisp morning air, enriched with the smells, sounds, and tastes of spring. In the meadow illuminated pockets of waving grasses reached for the rising sun. Buds of wildflowers struggled, hell-bent to breach into life and embrace the warmth of the sun.

"We don't appreciate the warmth of springtime without the cold harshness of winter to give it sweetness," he mused. The warmth of the rising sun coursing through his body brought a feeling that life was indeed a beautiful thing.

On this busy morning he listened to the insects drone, the wasps buzz, and watched the butterflies float and flutter, and the dragonflies instinctively hum through well-planned flight patterns to fulfill the measure of their creation and knit together the fabric of life.

The honeybees were first up each morning. Tom found these hard workers to be one of God's most inspirational creatures. They ventured out of their spring homes to mind these first flowers of the morning. Tom watched the journey of the honeybee as it followed a rivulet of sweet scent and pattern of color to the breathtakingly beautiful purple orchid and, now heavy with pollen, returned drowsily to its hive and queen bee.

He had read once that it's not love that holds the hive together, but the capable, disciplined, and determined sense of duty the queen bee has to keep the highly functioning family intact. Without her, their finely tuned operation would flounder. In a wistful smile a thought came to Tom. *The queen bee, a woman of substance.*

He lay back, put his hands behind his head, and looked up into the clear blue sky overhead, recalling another woman of substance he'd known. His thoughts drifted back to that long-ago morning when he'd referred to Annie as the queen bee of the butcher's shop. He smiled reflectively at the recollection. Over the years, she had become the heartbeat of that establishment, transforming it into the most successful in all of South Yorkshire. An amazing woman.

"Annie." Just the nostalgic thought of his friend brought a smile. She had grown up before his very eyes, and he hadn't even noticed.

Tom knew he must get on with the business of living. He could not go back to Woodhouse. There would be no work for him there. And the danger was too great for both he and his family. But he must move forward; he would go on to America.

Chapter 21

⋟ · ⋞

ON A WARM SUNDAY AFTERNOON in April, Annie lounged on the grass with Emma under the great spreading oak. Spring-time was everywhere. Together they watched butterflies flit from one flower to another. At twenty-one, Annie had grown into a tall, lean, and powerfully built young woman with intelligent dark eyes set in a clear, olive-skinned complexion. Leaning on one elbow, her long legs stretched out in front of her, Annie lazily picked the petals off a sweet-smelling daisy.

"You were in love with Thomas Wright, weren't you?" Emma asked.

Annie's head snapped around, but before she had a chance to object, Emma went on quickly.

"I saw it when you watched him leave Lydia's funeral last spring. It wasn't just in your eyes, everything about you testified to it. You were mesmerized. You didn't even notice me trying to get your attention."

"That's ridiculous." Annie's cheeks flushed scarlet. She didn't quite know how to protest. "That was a whole year ago, and you're bringing it up now? I wish you wouldn't do this."

But Annie remembered the moment. Before climbing into the carriage, Thomas had turned one last time to look back at what he had lost. The morning sunlight filtering through the oaks shone softly on the gravesite and the hundreds of beautifully placed flowers. As Tom turned back toward the carriage, his eyes caught Annie's. For just a brief moment, his gentle, mournful gaze had seemed to pass on an almost indiscernible expression of gratitude. Embarrassed for the selfish thought, Annie had lowered her gaze out of respect. But kindness is a language even the brokenhearted can surely feel. In that moment, Annie felt a bond sometimes forged between souls, often a result of unselfish, un-historic acts, quietly given of hidden lives in service to those they care about.

"I'm not sure why I've waited all this time to say something about it," Emma sighed. "I suppose I thought you'd trust me and bring it up yourself." Emma looked patiently into Annie's eyes. "I still see it in your eyes."

"Please, Emma, don't talk like that. He was my friend. He loved Lydia, and I was happy for him. Theirs was a match made in heaven. I was devastated when she died before they had a chance to realize their dreams."

"There was more than that, and still is."

Annie sighed. "I love you, Emma. But you and I grew up with seven cackling hens in a house with no roosters." She smiled, remembering. "We missed the perspective a good man can bring. What Thomas and I had was not romantic love." She paused to choose her words carefully. "Thomas was like an older brother to me when I desperately needed one."

Emma's eyes softened. "What do you mean?"

"People said he was a recluse. He had few friends, but Emma, he reached out to help me when I needed help. He mentored me, and I made him laugh." She remembered back, to the shy, insecure adolescent she once was. He had taken the time to engage her in conversation every time he came to the butcher's shop. She looked forward to those conversations, and over the years their relationship matured into a deep bond of friendship. "In some ways our friendship was a lot like yours and mine, but different. There was mutual trust, a profound respect for one another, empathy, understanding. I'll admit, there was shared affection between us, but it wasn't romantic." He had been interested in her well-being and she in his. They shared books, ideas, dreams for the future, and respect for one another's opinions. He quieted her fears, brought her through a difficult adolescence when she was confused, feeling alone. He buoyed her spirits, acknowledge her strengths, helped her become self-reliant, to gain self-respect as a woman. "*Friendship*—such a simple word. I suppose in some ways ours was a friendship that transcended age, class, and even romance. In some ways our unbreakable bond of friendship was more powerful than that of romantic love. Oh Emma, I don't really know? But I do know this—it's a friendship I miss."

"I'm so sorry, Annie."

"Don't be. He made me see things differently. He made me better, and for that I will always have a special place in my heart for him." Annie paused then gravely added, "But I'm no fool, Emma. Thomas had no interest in a romantic relationship with me. He was never interested . . . in that way! Lydia was his soulmate and always will be."

"Oh, Annie," Emma sighed.

Annie could see Emma wasn't convinced that this friendship of her soul hadn't connected irrevocably with her heart.

"Besides," Annie continued, "there's nothing to talk about, is there? He's gone. He's probably never coming back. Please, Emma, don't ever talk about him in that way again. There has never been, nor will there ever be, a romantic relationship with Thomas. I'm not pretty, vivacious, or elegant. I'm nothing like Lydia . . . and I never will be. Annie crushed the remains of the flower in her hand. "One thing I am, Emma, is a realist."

"So you've said." Emma looked back at her friend, seeing more than Annie was willing to admit, even to herself.

"We had great conversations, and I could make him laugh. I'm grateful for our friendship, but I'm sure Thomas Wright has never noticed me in that way. I mean *never*!"

Emma seemed lost in her own thoughts. "I don't know much about vivacious—and nothing at all about elegance, but you know what I think, Annie? I think men just want to be loved. They want to be told they mean more to you than anyone else in the world. And someone like Thomas Wright wants the companionship of an intelligent woman who doesn't use it as a club. When I talk to men at the dry goods store, a lot of them just seem lonely. I'm betting Thomas, wherever he is, is lonely right now." Emma raised her eyebrows and frowned. "Anyway, when I find a man like that, I'm gonna love him. And I'm gonna tell him so. It might take a while since I don't have any prospects, but that's what I think."

"What are you talking about?" Annie was indignant. "You're pretty and confident, Emma. You're like a magnet. They're always coming into the dry goods store for no other reason than just to look at you, to talk to you. Look at me. A strong-willed, quiet, plain girl without fortune. For goodness' sake, *I* wouldn't court me."

Emma frowned. "You're not a plain adolescent girl anymore, Annie. You've grown into an attractive woman—smart, the hardest worker I've ever known. Besides," Emma broke into muffled laughter, "you've become a wonderful cook."

"Oh, great. I'm sure that's a big selling point to a man like Thomas. A smart, hard-working cook he'd have to turn off the lights to go to bed with."

There was a long pregnant pause before they both burst into laughter. Annie wiped a tear of laughter from her eye. There was not much she could do to lighten her heart, but at least she could still laugh about it.

"I'm just glad I have you, Emma. Don't ever leave me. I can't bear the thought of you marrying one of those miners and leaving me all alone."

"No, Annie, you won't be alone. A man will love you because you're the kindest, smartest, most caring woman God ever blew breath into. He'll love you because you're a woman of great character, with the most prized gift God has to offer—the gift of humor. And unless I have misjudged you, you'll make him feel like one of those stallions. And that's a good thing. That's why a man will love you, Annie."

It was an honest response, and it lifted Annie's spirits a bit as she thought about the possibilities for her own future. "You are too much, Emma Stanley." Annie looked out onto the hills for a long moment. She sniffed and wiped at her eyes. "I miss him."

"I don't suppose we can choose the ones we fall in love with, can we?" said Emma with a dramatic sigh as she joined Annie looking to the horizon. "And I think it's a beautiful story of unrequited love."

"Oh, please! You can be so melodramatic. You should have become an actress. What would Grandma Nanny think?" But Annie had needed that encouragement. She didn't want to live a life defined by rejection, and Emma had helped her to see there were possibilities she had not considered. Maybe there was hope for love, even for a plain, bookish girl.

For Annie, confiding her innermost thoughts to Emma was like whispering into a cookie jar, then screwing the lid on tight, knowing her secrets would never be let out to hurt her. But even with Emma, Annie could not share some of her irrational, confusing, and most private feelings when it came to Thomas.

⊁ · ⊰

That night, Annie lay in her bed in the quiet darkness. The smells of curing meat hung just outside her door, wafting through her spartan furnishings. She was exhausted. She struggled to fall asleep. She could almost taste the dank, musty air that filled her nostrils, thickened with mold and mildew. In those moments waiting for sleep to come, after the barriers had fallen, disturbing thoughts of him drifted through her mind—his eyes, his voice, his expressive face, the soft wave of his hair, the easy way he moved his body. Despite her conscious pleas to turn away from these restless thoughts, she couldn't. Loneliness has a compass of its own. She tossed and turned, feeling restless, unsettled, like a bubble rising through water, only to burst at the surface out of breath.

It had been a year since she had last seen him, and she missed him. She breathed heavily, trying to clear the thoughts of him from her head.

Exasperated with herself, and unable to sleep, she lit the lamp and decided to try to read. The dim light drove the disturbing shadows into the corners of her room. Pulling down one of her many books from the shelves that lined the walls, she sought an escape from life's harsh realities. She tried to push the plaguing thoughts out of her mind with a book. But that night her books could not provide refuge for her tortured soul.

Finally, she gave up the effort to unburden her mind, to defeat the loneliness. She put the book aside, and again turned down the lamp. She lay for a long time in the quiet darkness, deep in thought.

She stretched her arms over her head and extended the taut thighs of her long legs out straight, then closed her eyes tight. She felt vulnerable, exposed. Her yearning mind drifting into private imaginings.

She wandered untethered into fitful dreams, lost in the warm glow of a multitude of beguiling desires, touches, and anxious caresses. It was heartrending, debilitating—her own private hell.

The morning half-light of a rising sun touched softly on the frosted windowpane, shattering its light into a rainbow of a thousand colors. Reluctantly she was drawn to the light that brought an end to her imaginings, back to reality. Away from the one-sided longing that gripped her chest, ending unwillingly with the pain of an aching heart, unsettled, lonely, heartbroken. Discontent with the knowledge her fantasies would never be, she chided herself for her foolishness, her reckless imprudence.

"It's all a dream. You must put these thoughts behind you. You are a fool in mourning for what you cannot have, nor ever will. You are weak. You are foolish. You are a silly creature. Let it go," she scolded herself. "Perhaps love is best left as a fallow field for me. I must move on. I can only get hurt by caring for someone from afar. Stealing love, but never sharing it."

Chapter 22

> ⇾ • ⇽

I<small>T WAS LATE SPRING, AND</small> the meadow was in full bloom when Tom made final preparations to leave his cavern home. He did one more check of his knapsack, filled to the brim with only the things that were most important to him. With a little tuck here, and a little tie there, he was ready to leave Spring Hollow as he'd found it and begin his life anew. Although this time, he had a bit more understanding of what was most important and a plan to make it happen.

He took a long last look at the pooling pond and estuary burgeoning with life. *Ripples*, he thought. *Grief has unending ripples of consequence, but some are for our good.*

Silently he took a moment to listen to the song of the lark bringing in the morning at the place he had called home for more than a year.

"Thank you," he said in farewell. "For not leaving me alone in my troubles—for giving me hope in my time of need."

As he began the long walk home, the warmth of the rising sun on this spring day struck the hillsides. From the ground millions of budding wildflowers sprung open to welcome the morning in vibrant colors of purple orchid, dark red helleborine, blue and red pansies, purple lilac, rock rose, the brightest white of spring cinquefoil he had ever seen, and the pale yellow-gold of primrose.

He drew in deep the sweet smell of the most beautiful face of springtime he had ever seen, a memory to be lodged forever in a special corner of his mind.

Some things are more precious, Tom thought, *because they don't last long.* He tried to resurrect Mam's favorite passage from the book of Job. *The waters of spring wear away the stones, washing away those things . . . intent upon destroying the hope of man.*

⊱ • ⊰

Thomas had called Annie his friend, but as she grew into her own woman, she had become more than that. There was nothing romantic about their relationship, but Tom couldn't deny that she had become special to him. On his frequent trips to Barnsley, he had often stopped by the butcher's shop, just to talk with her for a wonderfully invigorating moment or two.

Sometimes, it wasn't what Annie said, the empathy expressed, or even her growing knowledge on every subject imaginable that made her fascinating. She was different from him in many ways—their ages, their experiences, their outlook on life—and yet they found common ground. She had an enviable passion, an esteem for life, a compassionate heart, but more importantly it was what she did that set her character apart. The warmth that seemed to sweep into her heart when she spoke of Emma, family, and her empathetic responses to the small glimpses he gave her into his own family and relationship with Lydia. It was the hunger they both shared when they spoke of the freedoms America promised, the frontier, the wildlands where dreams could come true. He had been drawn to a certain something in the girl who had grown into an intelligent, fascinating young woman with a passion for learning. Theirs was a profound friendship difficult to describe, but nearly impossible to forget.

He wondered why human beings let physical attraction get in the way of an enduring friendship. He supposed an uncommon, abiding friendship like theirs might possibly have even more profound meaning without all the romantic entanglements. He wondered if the whole problem between men and women was that each might know what matters most along the long road of life, but still make poor choices because of confusion of the heart.

There was far more to Annie Dale than most ever saw. She would make a superb companion to face the trials, hardships, and demands awaiting him on the American frontier. She would appreciate the harsh, rugged beauty of it, and when times grew difficult, they could always laugh together. *So what if we don't love each other? Isn't it better to face the struggle with a lifelong friend—to share in the adventure with someone you respect and admire and enjoy spending time with?*

We could do it. I know we could. He pondered the thought. *I know this for sure: if I don't step up and take the future into my own hands, life will choose the course of it for me.*

He had given love a chance. He would not take that risk again. The consequences were too painful. But he knew if he were to get the full value out of life, he must share it with someone. With Annie— wicked smart, strong, reliable, an honest woman of substance with an inquisitive mind, interested in everything, and blessed with wonderful, self-effacing humor. He had known she was someone special from the first time they spoke. And though he had no illusions that she would ever be the great love of his life as Lydia had been, Annie had grown into an attractive young woman in her own right. A very capable queen bee. He knew what he wanted. He knew, if given a chance, their mutual admiration, shared values, could grow into something worth its weight in gold.

I can make her happy. And in time, we can have a family of our own to share in the bounties of America . . . if she's willing. His shoulders slumped and he exhaled through pursed lips, "If I can just convince her to take the risk."

On his long walk home, he resolved to do whatever was needed to convince this woman he admired and greatly respected to join him on his great adventure to America.

⊱ · ⊰

The dawn on the first day of summer was just peeking over the horizon. Annie, already busy with her chores, was sweeping and washing the front porch in preparation for the day. With the rising sun at her back, she stood quiet for a moment to admire her work. Her sleeves were rolled up, her arms wet, and her skirt tied behind her knees. Her wind-chapped face was further reddened from the work, and her unkempt hair loosely tied back in a ponytail.

"Hello, Annie," came a voice from behind.

Startled, Annie turned around to face the visitor whose voice she recognized but couldn't place. She leaned the mop against her shoulder and raised her hand to form a tent over her eyes, shading them against the rising sun. Squinting, she peered up at the man sitting atop the wagon seat. As soon as she recognized who it was, her heart leapt into her throat.

"Oh, Thomas, I've been so worried about you. I've missed you so!" Annie blurted out. The moment she'd said it she wished she could take it back, but it was too late—her feelings were out there. Just being in his presence brought back that old uneasiness. Frozen in place, she could say no more.

An infectious smile stretched wide across his face. The intoxication of seeing him again after so long gave way to an uncomfortable awareness of just how she must look. She fumbled to tuck a cloud of unruly hair back behind her ear, wiped her wet hands on her apron, then nervously smoothed it out in a pathetic attempt to make a difference, but it was of no use. And there was really no need—Thomas didn't seem to notice.

Annie prided herself in her ability to read the eyes of those she met. Faces changed with time and life's toll, but eyes remain constant, a window to the soul. But on this morning, she was flustered. She simply could not read Thomas's eyes beyond the reflection of some powerful thing afoot. What it was, she could not decipher. It wasn't disapproval, nor horror at her appearance, nor any of the sentiments she might have expected. There was still the inherent kindness in his eyes that she so adored. The warmth that, as usual, spread to the rest of his face. The remnants of the mischievous boy within this man were there. It all revealed a certain something about him that always seemed to send a tingling through her, down her spine to her toes. But this morning there was more. He looked at her fixedly, with a peculiar smile, and she could not discern his intentions.

Annie's eyes locked onto his as he stepped down from his wagon and calmly strode over to where she stood atop the steps. His shirt stretched tight over his powerful chest and shoulders. There was a healthy radiance in his rugged, tanned face and arms, his hair thick and dusted with chestnut from the sun.

Anxious, Annie's forehead began to sweat. The tension in her heightened even more at his quiet approach to the base of the porch steps coming up from street level.

"I hope I'm not too late, Annie," he whispered, pausing to wait for a response, but not taking his eyes off her.

Too late for what? she wondered. Her confusion was disconcerting.

Inhaling a deep ragged breath, he began to explain, "I've always felt you and I have had a special something between us. Our conversations outside the butcher's shop have made me a better man."

Annie stood quiet, puzzled.

He went on. "If I have misunderstood your impressions of me, please, one word from you, and I will make no more mention of it."

Now, Annie hung on his every word.

"But if I'm right . . . I am asking you to come with me to America?" He paused once again, brows raised, his hopeful eyes locked fixedly onto hers.

Stunned, overcome by his presence, and still trying to fathom the anxious moment, Annie was lost in his dark, smoldering eyes. She was quite undone. She almost didn't notice the bouquet of flowers he held out to her. Finally, in a daze, Annie reached her hands out to accept them. She stared down at the beautiful bouquet, so carefully wrapped to protect their delicate yellow-gold petals encircled in white.

"Primrose." She looked up into his deep brown eyes with flecks of luminescent green. "They're my favorite."

"I know."

"You do?" A look of wonder in her eyes, she stood deathly quiet, the mop still leaning against her shoulder. To beg more time to comprehend the full meaning of what she was being asked, she leaned down to draw in the sweet fragrance of the beautiful bouquet of primrose. The magic of flowers. Is it any surprise they are the universal symbol of life, death, and new beginnings?

Primrose. Her favorite . . . and he knew it?

The words of the question seemed clear enough and certainly straightforward enough. Even the dullest of girls could pick up on the words. But their full meaning she found difficult to comprehend, and the *why* impossible to fathom. She felt like a rabbit caught in a hunting dog's jaws. The more she squirmed and said nothing, the tighter those jaws fixed around her.

Intent upon uncovering the mystery, Annie locked her eyes onto his. "I'm not sure I understand just what you're asking?"

Tom smiled, sending a tremor of emotion through her. It seemed to Annie that he could sense all that was in her heart and see all that was in her thoughts.

"No, Annie, I've not lost my mind. I think I've found it," Tom's eyes beseeched tenderly. "I don't want to waste any more of my life being alone. I'd like you to be a part of it." He paused, looking deep into her eyes. "I'll promise you this: I will do my very best to make you happy."

It took just five seconds for Annie, the careful, prudent, pragmatic realist, to do the inexplicable—long enough for her hand to raise and cover her mouth, the mop to fall to the ground, and the strands of hair to pull free from behind her ear. Just that long to collect her thoughts—to know her answer. She must join her destiny with this man. She did not know all that he was asking, but she knew enough. She knew she loved him and supposed she always had. She was certain her life would be better with him than at the butcher's shop. He was a good man, and if it was a future in America he wanted, well then, she would go with him to America. Frankly, she would go with him wherever he wanted to go. If it was to hell and back, she was willing!

As Annie looked at him standing there in nervous anticipation, her words wrestled their way hell-bent through her bashfulness. "I will go with you, Thomas Wright. When do you want me?"

His eyes lit up. A contagious smiled swept wide across his face, bringing moisture to her eyes. He let out the long breath he had been holding and reached up to take her hand. "Oh, Annie, I promise I'll make you happy."

In an attempt to hold in her speechless emotion, her free hand covered her frozen smile, leaving the dimples rising on her blushing cheeks exposed for Tom to relish.

After a moment of shared smiles, Tom stepped back. "Do you mind if I pick up a few things in town, then come back for you as soon as I finish?" He paused for a response, and she nodded. "Okay then, I'll be back before dusk. Is that enough time for you to change, pack your things, and be ready to go?"

"Pack?" Annie blushed. "That won't take long. I'll be waiting for you here when you return."

Thomas released her hand and left the porch to climb up onto his wagon. He slid his rifle aside and sat for a long moment with the reins in his hand, looking down at her still standing there. His most elated smile filled her with warmth. It left her no choice but to return the smile.

As he snapped the reins in command, the horses moved into action. "I'll be back as soon as I can. Don't you doubt it for a minute, Annie Dale!" And off he went.

As soon as he was gone, the storekeeper's wife ran out on the porch all aflutter to alert the butcher's wife. "He's coming to take Annie," she shouted out, "and he's got a gun!"

Mrs. Jones, not yet fully dressed, stumbled through the door in a tizzy. "Annie!" she shouted in her foghorn voice. "You get back inside this shop this instant! He's a stranger to you, and in all proper decency, he should be acting like a stranger."

"I will this time, but this will be the last time," Annie snapped, still staring in disbelief at Thomas traveling down the road in his wagon. "When he comes for me, I'm never coming back."

><>•<

Annie sat next to Emma on the porch steps in front of the butcher's shop, anxiously awaiting Thomas's return. All her worldly possessions were held in the small satchel beside her. She had related every word of her conversation with Thomas, committed to memory like it was the catechism to the Church of England.

"I still remember the jolt of electricity I felt when he accidentally touched my hand that first time," Annie recalled. "I denied it, of course. But the truth is, from the moment he rescued me, my heart was irrevocably stolen and there was nothing I could do about it. How does one choose to deny the heart when it has such a powerful mind of its own?"

"Oh, Annie." Emma put her hand over her mouth, her brows raised in wonder.

Annie took a deep breath. "I'm sorry I never told you. But I was convinced it was a young girl's foolish infatuation. Then, I thought he'd gone forever. I was such a mess. Now here we are. I feel as giddy as a schoolgirl."

"Of course, I knew," Emma smiled. "It broke my heart to see you like that, but all's well that ends well, right?"

"I really don't know what to think." Annie grinned at Emma. "What am I supposed to think?"

"I wouldn't be thinking about missed conversations, that's for sure." Emma cocked one brow over squinted eyes. "There are more pressing issues, now that you have accepted his offer to go off with him," Emma suggested with a mischievous smile.

"Oh, Emma, what am I gonna do about *that*?" Annie put her forehead in her hands, wild imaginations racing through her mind. "How will I even know when he's making advances?"

With a wicked smile, Emma touched Annie's hand support-ively. "Oh, honey, you'll know. You well know the facts of life, don't you?"

"Hmm." Annie's mind ran on. "It's not the knowing that concerns me," she said with chagrin. "It's what to do with those facts I'm afraid about. I thought to catch a man properly, I was supposed to resist his advances first, before I blocked his retreat," Annie reminded Emma of their girlhood conversations behind the shed. "It seems I skipped right past all the blocking and parrying, went right to the . . ." Annie trailed off. "Will I be penalized for breakin' the rules?" The indiges-tion in her stomach was now rising into her throat.

"Oh, Annie. You're too funny," Emma giggled.

"Did he actually ask me to marry him—really? Maybe he's intent on makin' me his concubine, haulin' me off to some bell tower." She paused. "I don't know about these things, Emma. Except what I've read in those penny novelettes. Oh, Emma, I don't know." Annie dropped her head into her hands. "I'm afraid this could end in the kind of disaster where only the cockroaches survive."

Emma struggled to stop from giggling outright. "That's not what you should be thinking about," Emma tried to reassure her. "Think about the great adventure you're about to embark upon! You're leaving with a man you've secretly loved since you were but a girl!"

"I suppose I have to strike while he's still a bit crazy from living alone in that wilderness, before someone knocks some sense into him and he changes his mind," Annie rationalized. "Without any of that courting fanfare."

Emma tried to stop the laughter from escaping, but it came any-way. "This has to be the fastest courting of a man on record," she teased, doing her best to suppress the smiles. "Maybe you're just better at it than the rest of us?"

"Humph, I wouldn't bet on that." Annie frowned, pointing toward a pile of rocks. "Better off betting on the solidarity of them rocks in an earthquake."

Emma was still filled with a residue of laughter from the entertaining morning. "I'm gonna miss you!"

"I'll miss you too." The two sat together holding hands in silence for a long time.

Finally, Annie spoke. "Where is he? Do you think he changed his mind? I know I would have if I were him."

Emma elbowed her. "Not a chance!"

Annie tucked her strands of unruly hair back behind her ear. "He told me he'd be back by the time I packed and changed. Maybe he's expecting me to change into someone else?" The words flooded out in gallows laughter. "If he's thinkin' I might change into a woman that looks like Lydia, we'll all be dead before that happens!"

☀ · ☀

The warm glow of dusk was upon them when Tom finally pulled up in front of the butcher's shop full of apologies.

"I'm so sorry, Annie, it took a lot longer than I thought to fix the wheel at the smithy. Forgive me?"

Without a word, Annie stood up. Staring straight ahead without expression, she tried her best to withhold the flood of emotions. She stepped forward, unsteady, stone-faced. The tumult rumbling inside her head and churning in her queasy stomach left her weak, but still she plunged forward. In her left hand, she held tight to the handle of the small satchel, packed with the few things she could call her own. In her right, she held Emma's hand, who stood resolutely beside her. Annie felt comfort in the warmth of Emma's touch and turned to look at her one last time, her eyes brimming, her heart pounding, and her hands trembling. In a single glance, a shared message of love passed between them. Emma nodded in support. Annie's heart tore, for in this thing they could not be together.

Drawing on her shuddering confidence, Annie willed her buckling knees to hold her upright. She released Emma's hand and moved toward the wagon. It took all the determination and self-control she could muster just to keep from collapsing into a blithering jumble of nerves. But she was determined to control the volcanic eruption of emotions welling up inside her.

Just then, Mrs. Jones stepped through the front door, shrieking out her parting condemnation. "You're a damn fool, Annie Dale! Why are you going with that hermit? He's mad as a hatter. He'll probably shoot you with his gun!"

Annie did not turn back but responded looking straight ahead toward Thomas. "Because I love him," she answered, in as clear and resolute a voice she could muster. "Because I will always love him!"

"Well then," Mrs. Jones blustered, "you're fired!"

That might have been the parting benediction with which Annie left the butcher's shop had it not been for Emma, her anchor, her soulmate, who broke away from the gathered onlookers before Annie reached the wagon. Emma gave Annie a great comforting hug and, upon releasing her, took the bracelet off her wrist and put in on Annie's. The bracelet had been a parting gift from Emma's mam Hanna, when Emma and Nanny had come to live with Annie's family. As long as Annie had known her, it had never left her wrist.

"This is for you," Emma whispered. "It's for your life as Mrs. Thomas Wright. I will miss you desperately!"

"Emma Stanley," the butcher's wife screeched. "Come here! You're as bad as she is."

Ignoring the command, Annie's tall, pretty cousin clasped Annie in her arms again, kissing her fervently on the cheek and slipping a small bag of carefully saved coins into Annie's pocket.

"You're escaping for all of us," she whispered. When Annie tried to return the precious coins, Emma clasped both her hands over Annie's and kissed her yet again, whispering in her ear, "A wife must have a little money of her own. Remember this: you will always deserve the best if you always expect the best." With a final parting glance to her fondest friend, Emma wiped at her nose with the back of her hand. "Now get outta here and have a good life."

Clutching the few shillings in her right hand and the handle of her small bag in her left, which was adorned with Emma's bracelet, Annie Dale marshaled her remaining courage. Her dignity intact, she walked purposefully up to the wagon, though on wobbly legs. Thomas reached out a hand to help her into the seat beside him.

"You're quite something. You know that, don't you?" He smiled down at her.

Annie, cognizant of his encouragement, tried her best to offer a smile in return. It was a wistful thing, but a smile nonetheless. She took his hand and settled in beside him.

When a man loves a woman, he gives up little of his life. But Annie, in agreeing to be with this man, had given up everything. Her destiny was now inextricably tied to his.

Chapter 23

＞ • ＜

TOM SNAPPED THE REINS. WITHOUT another word, he and Annie drove off together into the gathering night. Annie sat next to him awkwardly, separated only by the rifle between them. Her anxious stomach was lodged firmly in her throat. She looked neither to her left nor to her right, but unflinchingly straight ahead toward her uncertain future.

Annie had always thought of herself as cautious and took no action without careful consideration. Yet here she sat beside a man she barely knew on their way to a family of coal miners she knew nothing of. They would be starting a new life on the other side of the world, Thomas had said. But America was a big place, and she had not asked where they would be going, when they might leave, nor what their new life would be when they got there. She just sat in ignorance, hope, and wonderment, bouncing down a country road.

"I'm not going to worry about it tonight," she whispered to the full moon. "Maybe tomorrow."

"Whoa, Blackie. Whoa, Sadie." Thomas pulled up on the reins, then turned toward Annie as the wagon slowed. "Shall we camp here for the night? Then we can try to get an early start in the morning. All right with you?"

Without a word, Annie gave him a shy smile in the affirmative. Her mind was racing. The inevitable night alone together had come. It was something she had dreamt of in her most private moments, on those nights lying in bed, unable to read, when barriers to rational thought had fallen before sleep came. What a twist of fate. She felt her resolve to do the right thing weakening as she sat quietly, watching him tying off the reins to the wagon brake and preparing to get down.

Annie grabbed the bedding and jumped down from the wagon seat. "No!" she blurted out in a shout of distress. Her heart pounding in a tumult of emotions, with trembling, balled-up fists, she strode off into the night in long, determined strides.

Tom jolted up in reaction. "Annie, hold on a minute." His eyes were wide in surprise. Then, he must have realized why she had reacted that way. "Fair enough, but here let me . . . you sleep in the wagon."

But Annie, of a practical mind, had already thrown her bedding under the spreading oak.

"You need to protect the wagon, the horses, and all our things. I will sleep under the tree tonight," she announced in a raw, raspy voice. Her feet spread wide apart, her eyes darting from Thomas to the wagon, she held her hand up. "Don't try to stop me," she cautioned.

"Okay, Annie. I'm not planning on it," he said, retreating, his hands up in surrender.

Annie, standing by her bedroll, took a labored breath. Her heart was pounding. Trying her best to be strong, she turned around to face him with her hands on her hips. "I have only one request, Thomas Wright," she barked out to disguise her unsteady nerves. "I want to be married proper. I want no bastards."

Seeming a bit flustered at her attitude, he called out from the darkness, "Goodnight, Annie. I'll see you in the morning."

With tears in her eyes, she took a shaky breath and whispered imperceptibly into the darkness, "Why am I crying?" Angrily, she wiped at her eyes. "You silly, prudish girl."

Disappointed in herself and reeling in emotions, she lay under the great spreading oak, staring up at the stars shining like sapphires in the night sky.

"Annie," Tom called out into the darkness. "Would it be okay if I took you to St. George's Church in the morning, where we can have the pastor marry us proper?"

She was quiet for a long moment, then wiped at her eyes and her runny nose. "Thank you, Thomas." A warm comfort enveloped her. Then for her own ears only, she whispered into the darkness, "I love you, Thomas Wright. Tomorrow night will be different. I promise."

It took a long time for sleep to come—for both of them.

⇀ • ↽

The minister at St. George Church was hesitant to marry them. After all, they did not yet have a proper marriage license, which included a waiting period. His hesitation continued until his wife, after hearing their story, pushed him into action.

"Oh, Herbert. Get off your high horse and marry this young couple. They can't sit around here and wait while you fiddle with it. After all, you're not vying for a position in the Trinity," she chastened.

Reluctantly, at the behest of his wife, the respectable minister agreed to perform the ceremony and she was the witness. Using the money Emma had given her, Annie paid for his services and a little extra for a marriage certificate. It was now official in the eyes of God and man.

Their courtship, engagement, and marriage had come and gone in twenty-four hours, but that day would change the direction of their lives forever.

⤜ • ⤛

The detour to St. George's had added yet another day and a half to their drive. Annie and Tom had begun on busy roads with the heavy traffic of coal carts, wagons, and transporters of commerce. But by evening of the second day, they were out of the busy villages, traveling on rutted and often muddy country roads, making travel slower. The languid pace on country lanes, canopied by spreading maples and fruitless mulberry, gave them an opportunity to talk. They spoke of many things, but their conversations always seemed to come back to America.

"As you well know, America has always been the place I've dreamed of going someday. My plans have been delayed a bit." Thomas looked at Annie, not wanting to burden her with his thoughts of Lydia.

Of course, Annie was well aware of the great love Thomas and Lydia had shared. He had looked at Lydia the way every woman wished a man would someday look at her. It had touched something in Annie's heart. She knew she would never be able to live up to their great love story, tragically cut short.

"I understand," was all Annie said. By unexpressed mutual consent, this hidden undercurrent of tension would remain a taboo, an unspoken conversation of the heart. At least for now.

"I'm not sure where I might find work, but I know for sure it will not be in Elsecar. I've burned that bridge. I'd be lucky if I didn't end up dead in a ditch."

She looked back at him, wide-eyed and appalled.

"Sorry, Annie." Seeing the concern in her eyes, he added, "I left on bad terms." After a pause, Tom continued. "Regardless, I will need to work long hours someplace to save enough money for us to make the trip. It's quite expensive." He paused to let his words sink in. Trying to elicit a response from Annie, Tom added, "Is that all right with you?"

"I'd say you've married the right woman. 'Cause I can work!" Annie exclaimed, trying to push back her anxiety. Then running on nervously, she blurted out in a rush, "God only knows I have spent enough time scrubbin' floors, cleanin', butcherin', cookin', totin' beef carcasses, luggin' furniture, and everything else imaginable."

"Well, all right then!" His smile grew even wider as he looked at her, grateful for her candor. "I appreciate that. I really do!"

She caught herself. Trying to quell her excitement, she looked down at her feet, hoping to find a response that would assuage her embarrassment. But she found no words there, so she pulled herself together and looked him in the eye.

"I'm sorry, Thomas. I guess I am just a little nervous." She felt her cheeks flush. "I just want to help you reach your dream—our dream."

Tom offered a grateful smile. "I'm thrilled you want to pitch in. It means a lot to me, especially after springing this on you with no warning. I suppose it was unfair of me," he reasoned, then added, "I'm not sure where I'll find work. We may need to leave South Yorkshire altogether."

"That works for me too."

This understanding seemed to break the ice and brought a warm, relaxed ease back to their conversation.

"I think you and I are going to get to know a different side of each other before this trip is over." His brows came together in a kind-hearted smile. "You'll let me know, won't you, if I say or do something that doesn't sit well with you?"

"I'll let you know," she smiled with increasing confidence. She felt safe with him. "Thank you, Thomas."

"Certainly. We are husband and wife, after all." He smiled at her. "What are your thoughts about this grand plan to go to America?"

"It's the land of opportunity, but there's a whole lot of it. A thousand miles between the northern and southern border, and more than three thousand miles from the Atlantic Ocean to the Pacific?" Annie saw the surprised look on Thomas's face. "That's a pretty big continent."

Tom shook his head, impressed by her knowledge. "That makes South Yorkshire seem pretty small potatoes by comparison, doesn't it?"

"Yeah. And I was hopin' you might be tellin' me where in those three million square miles you might be takin' me, since, as you said, we're married and all."

"What do you think about a farm out in the mountains of California? Maybe with chickens, cows, and pigs. We can grow some potatoes, corn, hay, and who knows what else?"

"I like cows . . . and pigs would be good," she acknowledged with a mischievous smile.

"You like cows, huh? Well, we'll have a whole herd of them, then."

She laughed. "I'm thinkin' I'd like the adventure of livin' out west in the mountains."

Tom tugged on the reins and called out again to the horses. "Well, Annie, it's been a long day. We're both tired, and I'm thinking this meadow might be a good place to camp for the night." He paused then turned to her. "What do you think?"

"Fine." She nodded.

Tom pulled the wagon to a stop, unhitched the horses, patted them on the rump, and sent them out to graze. He began to make camp alongside the stream, fussing over several seemingly unnecessary chores.

Annie thought she knew what was bothering him, especially after her antics of the night before. She was certainly not immune to those anxious feelings on their first night sleeping together either. Still, she said nothing.

She went about her own work, trying her best to look like she wasn't paying attention to him going from here to there and back again. She arranged their belongings in the wagon to make space on the gently sloping floor for their first bed, their feet at the low part in the middle and their heads at the raised part aft.

As dusk began to settle into night on this beautiful early summer evening, a full blood moon rose over the eastern foothills, and she began to worry there was something more on his mind. Tom attempted to tie the horses to the tree, but they were skittish and pulled away. Watching the horses fuss, Annie recalled something her father once said. He had worked in the Woodhouse stables and told her as a young girl, "A horse is like a mirror, it reflects your fears. If your horse is skittish and refuses to do as you ask, it is you who are not ready to face your fears." She frowned at the thought. The horses could sense Tom's uneasiness.

Annie got out a pan and unwrapped the cool cloth on the steak, a last-minute going away present clandestinely slipped to her by the butcher when his wife wasn't looking. Annie stacked rocks in a ring, then noticing that Thomas had not chopped firewood, calmly asked, "Where's the wood for the fire?"

"I'm not hungry," he responded impatiently, breathing heavily, his eyebrows knitted together over clenched teeth.

So, without fanfare, she repacked the cooking gear and returned the steak to its damp cloth. She knew he must be starving after the long day's work but could guess why he wasn't hungry. Even more concerned, she took a patient breath in the fading light of the evening and did her best to pretend to sew. Maybe he didn't want to be with her in that way?

Returning to the wagon, Thomas criticized, "You'll ruin your eyes." There was such solemnity in his voice that she almost burst into tears.

"Oh, Thomas," she whispered in a low, apologetic voice. "I so wish I were prettier for you."

He turned to face her. "Whoever led you to believe that should be horsewhipped, then shot! You are a perfectly lovely woman . . . in so many ways," he confessed in harsh tones, his voice shaking. "Don't ever say a thing like that again."

Startled by his sharpness, she stepped back, but he put his warm hands on her shoulders, gently reassuring her. She let him pull her close, and when their bodies almost imperceptibly touched, her breathing stopped. His dark, passionate eyes looked deeply into hers. A tremor pulsed through her body. She found it even more difficult to breathe. The soothing gentleness in his touch, the smell of his manliness, the ragged edge of passion in his voice brought a turbulent warming to her very center.

A warmth spread indiscriminately through her entire body. She could not have willed it otherwise if she had wanted to. Now her breathing came fast and irregular, her heart pounding. It was a feeling she liked feeling very much.

Then, he kissed her, and without warning she felt herself falling, then flying, her mind swirling into someplace she had dreamt of. She was lost in the warm glow of beguiling desire, and she began to cry. The tears coursed down her cheeks. With desire raging through every part of her body, she shyly awaited his next advance. But it didn't come.

Tom pulled back. "Annie, are you all right?" he paused, waiting for her to respond. "I'm so sorry! I didn't mean to . . ."

Her head was spinning. She felt unbalanced by his touch, could not breathe, and was unable to find her voice.

She sensed the urgency in him. He was clearly trying to get control of his emotions. In his confusion he opened space between them, his trembling hands still on her shoulders. Desperately, she wanted the clean lines of his lips and the warmth of his mouth on hers, but her shyness won out. She blushed and looked away.

"Tell me that you don't want me to kiss you, Annie," he said in a hesitating voice. "Tell me you don't want me to make love to you."

"I can't," she whispered, returning the intensity of his gaze. "I can't tell you that!"

Annie could see from the brightness of his eyes and the way he drew in his ragged breath that he wanted her. And she knew she desperately wanted him. Flooded with desire, she could wait no longer.

She put her hands on his chest, and reaching up on her toes, she kissed him passionately. His lips were surprisingly soft, supple, and firm all at once. The taut muscles of his chest, the sound of his pounding heart, and his masculine smell were intoxicating.

He kissed her back, further coaxing her mouth open with a soft persistence.

A wonderful sensation traveled through her whole body. He wrapped his arm around her waist and pulled the full length of her body tight up against him, his other hand intertwined in her hair. She pushed herself into him, and as she did, a little sigh escaped him. Tilting her head to the side, her lips softened. As he kissed her again, soft and gentle, every nerve in her body was set to humming. It brought thoughts of her dreams, of illicit affairs of the heart.

He lifted her easily onto the wagon bed and laid her down gently on the soft bedding. Her breath came in short, sharp intervals and her wild eyes flashed to the night sky, strung with millions of stars above them. He leaned down and kissed her again, and again . . . and again and again. And with awkwardness, confusion, and the uncertain touch that comes with shared inexperience, Thomas Wright took Annie Dale to be his lawfully wedded wife.

Chapter 24

✦ · ✦

I HAD A LOT OF QUIET time in Spring Hollow to sort things out—learned a lot about the way things are, and a bit about myself too." They ambled down the roadway on the last leg of their journey to his family's old farmhouse. He turned and looked at Annie for a reflective moment. "As I recall, you consider yourself a practical realist, always careful and pragmatic. I can imagine you have a few questions you might like to ask me."

"Oh, yes. Careful and pragmatic." She rolled her eyes. "You mean like how I insisted taking an entire thirty seconds to consider my options before entering a lifelong commitment?"

"I admit, I was surprised by your quick decision." Tom shot her a warm smile.

Annie sat pondering for a moment, then said, "Since you're offering, I do have a couple of presumptuous questions." She paused, then asked, "What was it like living all alone in the wilderness?"

"I wondered when that question was coming," Tom grinned. "At first, I was consumed only with survival—a blessing, really. I grew to appreciate the discipline required to live in the wilderness. Out there, I could breathe, and there were no eyes on me but those of God."

Annie gazed ahead at their two horses plodding down the quiet road. "I'd be so lonely with no one to talk to."

Tom pursed his lips, considering the question imbedded in her comment. "I was lonely sometimes. But being alone in your thoughts has its benefits. For a long time, I didn't want to talk to anyone." Tom remembered how he learned to appreciate small and simple things, to listen to that still, small voice inside, to wonder why we're all here. "I spent more time considering what I wanted to do with the rest of my life."

"I suppose we don't often slow down enough to take that opportunity, do we? We're so quick to get on to the next thing, seldom appreciating the moment."

Tom nodded. "And I was no different. But I wasn't entirely alone. The animals of the glen became more than just a meal for my survival. On a certain level, I found them . . . essential to help keep my fragile sanity, which I'm not sure was entirely successful." He chuckled. "I came to look at things differently. I found there to be a certain sacredness in their lives."

"I appreciate animals as much as anyone," Annie said, "but have never considered that."

"Have I confirmed the butcher's wife's suspicions? Am I sounding like a crazy loon after all?"

"Ah, Mrs. Jones!" She laughed. "You're probably crazy as a loon, but that kinda works for me—fascinatingly crazy," Annie smiled wryly. "So, tell me, how did your time in the wilderness change you?"

Tom paused for a long moment. "When I was a boy, so many died of smallpox in our shantytown that Mam insisted we children be inoculated. I was horribly sick for three weeks from just the inoculation—I felt like I wanted to die."

"I had a similar experience. It was awful."

Tom nodded. "Surviving the Spring Hollow inoculation helped me see things differently." Tom remembered how difficult those first nights in the glen were. He learned that life is not made unbearable by our circumstances, the challenges we face, but by what we do with the pain from them. "Surviving heartache in the wilderness helped me order my mind, clear my focus on what to do with the rest of my life."

"It's a testament to your character that you were able to draw strength from your grief," Annie proffered.

"You know," said Tom, "I often reflected on our discussions in front of the butcher's shop when I was alone in my cavern home. Funny as it might sound, I missed them." He felt a warmth rise to the surface as he remembered those emotions. "I missed you."

Annie blushed. "You'd be surprised how funny it ain't," she said, smiling. She paused a moment, then meekly asked, "Is that what inspired you to ask me to marry you?"

A cautious smile turned up the corners of Tom's mouth. "Last autumn, I became fascinated by a beautiful young doe that frequented the pond every morning." Tom remembered her graceful features reflecting delicately off the mirrored surface of the pond, through the blue-grey mist drifting up through the canopy of the giant sycamore. "In the soft morning light, Elsa looked almost elegant."

Annie smiled, her eyes crinkling around the edges. "Elsa . . . she sounds beautiful."

"My name for her, and she was. By fall's end, Elsa had entered into courtship with a buck." Tom paused. "Did you know roe deer are monogamous?"

"No, aren't deer herd animals?"

"Roe deer are the only deer in Europe who mate for life."

"Really!"

"I suppose I was just lonely," Tom sheepishly qualified his words, then went on, "but I became obsessed waiting for Elsa and her beau to show up each morning."

"It seems we all need someone, don't we?"

Tom nodded. "While I was making preparations to leave Spring Hollow, Elsa and her buck returned one spring evening after wintering in the valley." Tom remembered how Elsa gracefully strode out of the thicket, across the meadow. Her buck beside her was striking— majestic, really. The setting sun reflecting off his new velvet-covered antlers. He seemed more protective, purposeful in his bearing, than he had in autumn. Tom remembered the stirring in his heart, feeling exhilaration deep in his soul. "There was something different in them. I couldn't put my finger on it." Tom paused, his elbows on his knees as he watched the two horses leisurely lope along down the roadway.

"Suddenly," he continued, "a rambunctious little fawn burst out of the thicket, hurtling himself after Elsa. A handsome little thing— soft red coat, refined spotted face, alert ears, and tentative legs still not quite sure of themselves."

"Oh, how lovely."

"He weaved in and out between his mam's legs. Then settling in, he pressed in close to Elsa, looking back toward the thicket at two very upset crows chasing after him."

"How fun! I would love to see something like that."

"Elsa turned to face me, paused, and looking in my direction with those enormous brown eyes, she caught me up in a feeling of wonder." He cautiously considering his next words. "I imagined we shared a message, she and I—a message about family." Thinking Annie might consider him of a disturbed mind, he qualified, "Probably just my lonely heart. Of course, there was really no love there"—he supposed Elsa was just fulfilling the measure of her creation—"but it seemed real enough to me then."

"Oh, Thomas, I can imagine what it must have felt like, but to be alone. . . I would have wanted to share it with someone."

Tom sat quiet, holding the reins as the horses plodded along. "I did want to share it with someone." He had remembered his many conversations with Annie and her fascination with wildlife. He'd imagined being with her and their own family, sharing experiences like that one, fending for themselves, beholden only to the laws of nature. "I wanted to share it with you, Annie." He caught hold of his emotion and reined it in. "Sorry—I suppose I'm sounding crazy again."

"I don't think it sounds crazy," she whispered, but could say no more.

Tom took her hand in his. "You'd be surprised how often during my isolation in Spring Hollow, something you'd said came back to me in a new light. You were so young when we met, but you grew into a fierce woman. And I hadn't even noticed it."

"Fierce?" Annie reacted with soft laughter. "Well, that's gotta be attractive."

"It is! Your passion for life's most important things opened my eyes to another point of view." He remembered how they had made him consider things he hadn't before on his blind march toward his goals. "I missed our conversations—our special friendship. It had grown to be an important part of me over the years."

Annie was silent.

He remembered the hope to take Annie with him to America further coalescing in his mind. "As I made the long walk back home to Elsecar, I knew if I could talk you into it, we could have a good life together. I was ready to get on with life, and I had some crazy hope you might just agree to be a part of it."

Annie sat quiet listening to Tom finish his story. "I suppose I am now a fan of being alone in the wilderness," she whispered. "I played 'hard to get,' didn't I—for about fifteen seconds."

Tom laughed. "You're a trusting soul." He gave Annie a long look. "I don't have any money, but I have picked up some infinitely precious experiences. They didn't come free, but unlike most all else in life, they can't be stolen. Those experiences have changed the direction of my life."

"Mine too!" Annie put her arm through his as dusk approached, then leaned her head against his shoulder. "Mine too, Thomas Wright."

❧ · ❧

Minutes passed without a word from either of them as Blackie and Sadie plodded along on the final stretch of the road to the farmhouse. Tom clicked the reins in encouragement and turned to her. "Annie, I know I'm taking you away from everything you've known. All I can offer is a hope of something better."

Annie grinned. "Right—taking me from the butcher's wife? I feel like a prisoner just sprung from the Clink! And, I suppose, if a little adventure is on the menu, I'll take some of that too." She hugged him.

Tom chuckled.

Annie was now more resolute than ever to make a success of this new life with her determined husband. It would be a hard life, of that she was sure, but hard work was one thing she was good at. Annie felt the warmth of emotion as she imagined her future. She was moving forward with a man she had always secretly loved. Well, maybe not so secretly. It felt good. *He may not really love me . . . yet,* she considered. *But in time, maybe?* It gave her a goal where there had been none before.

Now was not the time to discuss his feelings for Lydia—that time would come, but not today. "Have you heard anything from George?"

Tom shook his head. "Not a word since he left the *Chicago Tribune* heading west. I fear the worst. He may have lost his life in America.

We forget how brutal life is in the Wild West. But he lived it well—he was willing to risk it all in the hope of gaining it all." Tom looked at her with a comforting smile to ease the tension. "George wrote a phrase in one of his last letters: 'Fear is a reaction; courage a decision. And when courage is chosen, it's often contagious.'"

Annie smiled. "Wise words. Not hearing from him must be very hard."

"Devastating." Tom furrowed his brow. "I've promised Mam that I'll find out what happened to him. Or at least lay the mystery to rest." Tom paused on that thought. "You will love my mam, Martha. She's a lot like you. A voracious reader and interested in everything."

Annie turned to him and took a calming breath, a look of trepidation in her eyes.

He tried to be reassuring. "Don't worry, you'll love them."

"It's not them I'm worried about. I can only imagine their shock when I walk through their front door." Annie was apprehensive at meeting her new in-laws. "They will let me through the front door, won't they?"

"We're talkin' 'bout a house full of coal miners, Annie," Tom grinned. "We've got but one door. And until recently, only one room. And that's only because all in the family chipped in. We share a ramshackle farmhouse on the edge of town. It's chock-full to the brim with family. Mam's a force of nature. And you'll adore my sister Edie, her five little ones, and her husband Andrew. They're all stuffed into one of the two bedrooms. My parents live in the other room, and now my younger brother Joey sleeps in the front room. We may be in the barn with ole Blackie and Sadie for the next couple days. But trust me, they'll love you until you can't take it anymore."

Annie nodded. She still felt nervous to meet his family, but she hoped for the best.

"Should we even be going that close to Woodhouse, with Boo Black around?"

"We won't stay long. And I don't expect to go into the village."

They passed through Huddersfield, a tidy little village just a few miles southeasterly of their destination. Just beyond, as the sun was about set over the Pennines mountain range, they entered an isolated, charming meadow with splashes of wildflowers displayed like pools of color on a painter's pallet. Only the soothing sound of a stream broke the silence. There was a sweet-smelling crispness in the air as the sun slipped behind the mountains. A muted blaze of color reflected off the underside of scattered clouds, casting shimmering shadows on the swirling eddies of the brook.

Tom pulled up on the reins and looked around. "Well, does this look like a good place to spend the night?"

Smiling, Annie nodded. "Looks to me the perfect place to spend the night together!"

>· ◄

Things went differently when nightfall came on this, their final night on the road as husband and wife.

"Isn't it about time for bed, my husband?" Annie suggested as she sat in front of the fire; having finished their chores, both were anxious for their night together.

"Did you have something in mind?" Tom replied, with a gleam in his eye.

"Why, yes!" She looked up at him coyly as he stood beside the fire, tethering the harness. She felt the glow in her cheeks rise and stifled a smile as she shyly stood up and came to him. He put aside his work, and affectionately put his arms around her. Lifting his hand, he traced the lines of her face with his fingertips barely touching her pale skin in the soft light of the fire. She shuddered at his touch on her ears, her neck, her cheek. She could feel the tension in him, the questioning look in his eyes searching hers for permission. This time she didn't hesitate. She boldly held his gaze, wanting him, and knowing she had the flint to light the fire in him.

"This has been quite a week for you," he whispered. "You were fired, your first date, your first trip to Wentworth."

"And other firsts too," she whispered, her voice cracking.

"Your first kiss?"

Without a smile, she responded, "Already been there and beyond."

"So, what did you have in mind?"

"I'm thinkin' it's time for you to stop talkin', my husband!"

Not releasing his eyes from hers, he placed his left hand on the small of her back and slowly pulled her into his embrace. With his right hand, he threaded his fingers through her hair. Tingling with anticipation, she breathed in deep trying to settle her emotions. She could feel the passion running through him. She pressed her breasts against his chest. As he coaxed her mouth open with a soft persistence, a wonderful sensation traveled through her whole body, settling in her very center. She was not prepared for the intoxication of the moment, her breath coming fast, her blood rushing like a tide into every hidden crevice of her body with his magical touches and anxious caresses.

He pulled back and looked into her eyes. The beguiling, almost chilling intensity of his eyes on her held her spellbound. They seemed to bore right through her. She was sure he could hear her ragged breathing and sense her ardent, pulsing desire.

She wrapped her arms around his neck and kissed him long and hard. Passion surged through her as the length of their bodies pressed together.

"Oh, Thomas." The words escaped her lips in a whisper.

Her heart was pounding as he swept her up and placed her gently in the back of the wagon. Lying under the stars, wrapped in warm down covers, all the world around her was forgotten— all but his callused hands on her body, his rough cheeks and his warm mouth on hers. Her breaths came fast and halting. Her whole body trembled, throbbing in a deep yearning to fill the hollowness within. She wrapped her arms around his shoulders, curled her fingers around the taut muscles of his back, and with all her strength pulled him into her. A desire she had not even dreamt of swept through her, shaking her to her core as she was swallowed up in emotion. Her breathing stopped. The mystical intensity of their joining reached parts of her she never even knew existed until, finally, it was finished.

They lay exhausted in the nest of blankets and pillows, holding each other under the stars strung out like diamonds in the night sky. They pulled up the warm blankets to their chins until only their eyes were exposed. Sharing smiling glances, kisses between brief words about the beauty of the night, they spoke of their dreams for the future.

"Well . . . you promised a little excitement!" she said, feeling her cheeks glowing.

Tom chuckled. "I certainly think we've made a good start down that path, don't you?"

"Start!" She smiled back at him, her raised brows pulled together. "Wow, if this is the start, I can't wait 'til I reach the finish line!"

⇝ • ⇜

As the sunlight crested over the hills and into the valley the next morning, Annie lay in the back of the wagon watching Tom whistle while he worked.

"You're a morning person," she said, pulling the blankets up to her chin against the coolness of the dawn.

He turned and smiled at her. "I'm sorry for waking you." He handed her a cup of tea. "I'm making omelets. How hungry are you?"

"Hmm, I'm pretty satisfied," Annie teased as she accepted her cup of tea. "Thank you, my husband, for what you have already given me."

"No, my dear, it is you who has given me more than I deserve." He leaned over the bed of the wagon to give her a good morning kiss.

Her brows raised mischievously. "I always try to do my best at everything I do." She looked at him from her bed in amusement. "I see you've already eaten your omelet. Worked up quite an appetite last night, did ya?"

He laughed contentedly.

She would miss only a few things from her old life. The companionship of Emma would be the biggest, of course. Yet, she was surprised at how easy it had been to put thoughts of Emma away. She almost felt guilty. Her love for Thomas had only intensified these past few days. He had a way of making her feel like she could slide into his life easily. She couldn't wait for the next chapter.

After breakfast, Thomas refused to let her help him pack up their camp and hitch the horses. She took the opportunity to write a quick letter to Emma.

Dearest Emma,

Thank you for everything. We got married at St. George's, thanks to your wedding present. If I'm being perfectly honest, the groom was prettier than the bride.

The guesses you and I made under the spreading oak about the mysteries of men and women were so nearly right, our nuptials had a certain pastoral grandeur about them. It's not perplexing really! Men are simple creatures. It's all about food and . . . Well, let's just say I've learned a whole lot more about stallions. I'm not sure any of my organs wafted, but I don't expect a hysterical breakdown any time soon.

There is something to be said for the first love in a man's life. Lydia will forever hold the exalted pedestal of being the first in his heart. But maybe, just maybe, she will not be the last. He'll probably never again have another soulmate, but the distinction of being the last love of his life ain't bad either.

There is solace in knowing that I will always have you, Emma.

Love,

Annie

Chapter 25

$\succ \cdot \prec$

RIGHT FROM THE START, EVERYONE in the Wright family liked Annie. As they spent the day together getting to know each other, she was relieved to find they were both kind and accepting.

"I hate to say it, Tom," Joey said as they were finishing dinner, "but I've missed you and your curious ways."

"We all missed you," Martha added. "And we want as much time as possible with you and Annie before you whisk her across the Atlantic."

Tom cleared his throat and sat back in his chair. "Annie and I have really enjoyed spending this day with you. We hate to leave, but of course we can't stay here. With the bad blood between me and Boo Black, and . . . well, we all know how unforgiving Lord Fitzwilliam can be. Living and working around here puts not just me, but our whole family at risk."

"That's true," Papa chimed in.

"It takes a lot of money to go to America and make our way west—money we don't have. We were thinking maybe we might find work down south by Gleadless. We still have family there. Or even up north. There are a lot of coal mines in Wales."

Martha looked at the rest of the family surrounding the table. "We'd like to talk to you about that, Thomas."

"What?" Tom looked around the table at the telling faces of his family.

"While you were in Barnsley," Martha continued, "proposing to this young lady, we have been looking into other opportunities away from Elsecar and Woodhouse. But not too far away. Your brother has spoken with the folks at the Oaks Mine, and I hope you'll forgive me, Annie, but I have made inquiries about work for you with Thomas's cousin in Hoyle Mill. That's just four miles from us, seven from Elsecar, and a bit farther to Woodhouse and Lord Fitzwilliam and—"

"Tom," Joey cut in. "I've been testing the waters around this part of South Yorkshire, and everyone at the surrounding mines, including the Oaks, knows of your slight to Lord Fitzwilliam and the infamous encounter with the Candyman when you left."

"Oh really." Tom laughed in disbelief. "The rumors spread all the way to Hoyle Mill?"

"Of course, Mick told his mates what he'd witnessed. Then they told and retold the story with the requisite embellishments. You know how the mining community is, word spread like wildfire up and down the seam. Your reputation has grown in the retelling—'The Notorious Thomas Wright Affair,' they call it," Joey laughed. "Your standoff with Boo Black has become something of folklore—legend, even."

"Ridiculous. You're makin' this up Joey." Tom looked dubiously at his brother, then around the table at the rest of the family.

"It's true enough," Martha confirmed. "But don't make light of it, Joseph."

"There's talk down at the pub of the company's disappointment in Boo Black," Papa added from the other end of the table.

"Of course," Joey agreed, "Boo's not happy about it one bit, so ya got an enemy there for sure. But to the miners, well . . ."

"So, ole Boo's gonna club me to death on sight?" Thomas interrupted, "and throw my body into a ditch?"

Martha shuddered. "Please stop it, you two. That man's no one to be fooled with. While some may foolishly fear Boo Black less, Tom's humiliation of that brutish lunatic has made him even more dangerous to this family."

Papa interjected. "These are dangerous men. Boo Black will never forget or forgive."

"Sorry, Papa, Mam," Joey apologized.

"He may not own the mine," said Tom, "but Lord Fitzwilliam is the landlord for much of the buildings and property around the Oaks, isn't he?" Tom pointed out.

"I'm not sure how deep the connection runs between Lord Fitz and Thomas Dymond, the owner of the operating company for the Oaks Mine. These aristocrats all have incestuous relationships. But the Oaks desperately needs your expertise, Thomas. In the eyes of some important men your stock has risen." Joey looked at Annie. "It's all about money. The company is missing out on a lot of it, and Woodhouse is eight miles away. Enough distance to give Tom some breathing room from His Lordship and Boo Black." Joey paused, watching Annie bite her lower lip. She was clearly becoming anxious about the implications of Joey's intriguing train of thinking. "I've been laying some groundwork for my big brother with the Oaks colliery's new pit manager, Tutor Turton."

"Tutor Turton," Tom interjected with raised brows of surprise. "Mr. Prudent. How did that happen?"

"Things change. You left the mine and went to work for the Silkstone & Elsecar Coalowners Company in Woodhouse, and Tutor went to the Oaks. The Oaks has mechanical problems, operational problems, labor problems for years, and . . . well, all seems to have gotten worse this past year. Production has fallen while the demand for coal has expanded exponentially. Your skills are in high demand, Brother, and your reputation for production strategies precedes you."

"Hmm!" Tom's mind was swirling, trying to keep up with the meaning of all this.

"You know how these companies are when it comes to their financial bottom line. They'd trade their mother for a few shillings."

"What are they willing to do?"

"Well," Joey began, "Tutor is on trial, frankly. He tells me he's got maybe six months to turn things around. He knows you'd be a great asset, maybe even save his job. He's been lobbying for you, Tom. He's convinced the company that you may be the answer to their prayers to their money gods—who, in this case, seem more important than their arrogance. Tutor says they're convinced you may just be the keystone to solving their circle of financial woes."

"And how are they gonna protect my husband from this dangerous monster Boo Black?" Annie asked.

"Hmm." Tom stood and began pacing the room. "I can't imagine Boo would let this go. He's got too much arrogant pride to let it go. I'm sure he'd do anything to see my head on a platter."

"I'm sure both Thomas Dymond and Lord Fitz consider you an insolent thorn in their side," said Joey, soberly addressing his brother's concerns. "But both are businessman who stand to make a lot of money if the Oaks does well. And I suppose, being eight miles away from Woodhouse, you might be out of sight and out of Lord Fitz's mind. Tutor says he's secured an agreement to put a leash on the Candyman. Boo wouldn't dare cross Lord Fitz."

"Can we trust them?"

"I trust Tutor, of course, but who knows what clandestine intensions are afoot between Dymond and His Lordship."

"Oh, Thomas, I'm afraid for—" Annie began to interject.

Joey interrupted with an apologetic nod. "There will come a time when that leash will be taken off Boo, and he will want to exact his pound of flesh. But for now, it's all about powerful men and their insatiable greed. Hopefully, you and Annie will be long gone before the tide changes. You know how these aristocrats covet the almighty pound more than anything else on earth or in heaven above." Joey looked into Tom's eyes. "In part, thanks to Tutor, they're of the opinion your ingenuity could be the balm to heal the injury to their financial ledger."

Tom and Annie looked at each other silently for a long moment.

Martha broke the silence. "Not to influence you unduly, but I should tell you, Cousin Will's wife Emily has found Annie employment at the Hoyle Mill Mercantile. And she has offered to have you and Annie stay with her. She's pregnant, you know, and needs roommates to offset expenses for a while."

"Roommates? In that tiny place? How could Will and Emily squeeze in roommates?" Tom asked.

Martha sighed, her voice tinged with sadness. "Will's not there. He got on the wrong side of our illustrious Lordship, and he threw Will and also Emily's brother Mica Wilkinson into prison."

Joey interjected, "They'll be workin' a chain gang for the next three or four months."

"That's ridiculous. Mild-mannered Will? What for?" Tom asked, more than a bit disgusted.

"That's a discussion for another time, Thomas," Martha pushed on. "The point is, Emily needs some help. And now that I've met Annie, I know she'd be a perfect fit."

Tom turned to Annie. "We're in this together now. I won't consider it if I don't have your support. This opportunity at the Oaks and Hoyle Mill will make us more money faster and cut our expenses. We can leave for America sooner, but there is danger in it. What do you think?"

"I like the idea of being close to your family—my family now— but the risks from that awful creature terrify me." Annie looked at her husband with worry in her eyes. The rest at the table looked at her for an answer. She swallowed hard. "I'll support whatever you decide."

Tom registered Annie's lukewarm support. "Shall we see what Tutor has to offer in the morning?"

She nodded her reluctant affirmation.

"If you do agree to take work at the Oaks," Martha added, "You must stay away from Elsecar and Woodhouse at all costs."

>•<

The next morning, with Joey and Andrew by his side, Tom calmly strode through the Oaks Mine's iron gates.

Boo Black and his gang of thugs were waiting there to greet him.

Boo stared at Tom long and hard. "Enjoy it while ya can, 'cause in due time yer mine," he spit out with a smirk.

Tom's stand against this hated thug had stiffened the spines of his fellow collies. Several miners moved in around Tom, glaring at a diminished Candyman. They quietly let their support for Tom be known, just in case things got out of hand with the company enforcement brigade.

Tom straightened to his full height. His piercing brown eyes shone with quiet but fierce confidence. "I'm here to meet with Tutor Turton." He raised both hands in reconciliation to ease the tension. "I want no trouble." With that, Tom walked on in to see pit manager Tutor Turton, accompanied by a large contingent of miners.

>•<

"Hello, Tom! It's been a long time."

Tutor welcomed him into his office. He had grown into a tall, sallow-faced man. Just three years older than Tom, he was still "Mr. Prudent," as Tigger once described him, careful in his approach to situations that needed proper handling and judicious in carrying out solutions. Tutor pulled off his hat, placed it on the clean desk in front of him, then reached out to shake Tom's hand. "What can we do for ya, Tom?" he asked, knowing full well why Tom was there.

"I've come looking for work, Tutor."

Tutor nodded. "Ya knows how I feel about ya and your family. Fifth-generation workin' man along this here seam, but ya made quite a reputation for yerself." Nodding toward the gang of miners standing outside his door, Tutor anxiously swept a hand through his hair. Both knew they were being watched, and one unplanned move could upset the apple cart.

"Oh, I wouldn't believe everything you hear, Tutor," Tom responded without a readable expression. "You know how tall tales are, they can travel all the way to London and back before the truth gets outta bed and puts its boots on."

Tutor forced an uneasy smile. "Maybe so. But ya upset a lot of folks up at the mansion house when ya left." He paused. "Now I know Boo Black overstepped his bounds, but I suppose ya knows how things are, and Lord Fitz . . . well, he has a long memory."

"I have no interest in causing you or the company trouble with Lord Fitz or anyone else, Tutor. I just need the work."

"Joey mentioned it, but the company folks is hard struck. And the miners sittin' out there have made it clear they want no monkey business. That's a bad mix." Tutor paused to let his statement sink in, then looked straight into Tom's eyes.

Tom calmly returned the gaze but said nothing more.

"Here's what them company fellas told me I can do. They'll hire ya—but no leadership role with the men." Tutor showed Tom a piece of paper with the amount they'd pay. "No bonuses, and you'll have to sideline any of your special safety projects." Tutor scratched his chin as he waited for an answer. When he didn't get one, he continued, "O' course they'll be wantin' to use your production and technical skills. That's what I can offer right now." There was a long pause of uncomfortable silence. "They need ya here, Tom . . . badly. I'd deny I told ya so, if anyone asked, but it's true."

Tom nodded, satisfied. "All right, Tutor."

Tutor's shoulders relaxed. The color came back to his face. He wiped the perspiration from his brow with his sleeve.

This was not Tom's boyhood friend speaking. The company made all the rules here, and every man either played by them, or they were drummed out of the valley. Or worse.

"It goes without sayin' there'll be no more controversy. And ya'll speak to dem fellas sittin' out there, tellin' 'em ya been treated fairly and want no trouble. Of course, there'll be no mention of your wages by either you or the company. Do ya think ya can do that?"

"Thanks for the opportunity, Tutor." Tom looked him straight in the eye. "I can do that."

Tutor let out a long-held breath. "Tom, I hope ya know I hates this part of the job. Ya have always been a man of your word, and I takes ya at it! And after Fidget's death, you and your mam done right by my family. I'll be forever grateful for that."

The two men looked at one another with mutual respect, and the deal was done with a handshake.

"The men all have great respect for ya, including most in the company, at least for your abilities," Tutor said. "Of course, ya know how I feel about ya. What ya done with Boo? Well, that was for all of us workin' men."

"You better watch yourself, Tutor. You may be in danger of becoming a good man!"

Color came to Tutor's face at the compliment. "Here's hopin' it'll pass quickly." Tutor tipped his hat in respect. "Give my best to your family. I'll be a seein' ya Monday morning then, fair enough?"

"Fair enough, Tutor. Say hello to Mary and the kids for me."

"Tom," Tutor said, dropping his voice low. "Please be careful. Try not to be alone, especially at night walking home. There are those who are waitin' for an opportunity to do you harm. Boo's chompin' at the bit and will be lookin' for any chance to take ya down." Tutor paused. "He may seem far away at Wentworth, but ya know what he's capable of if he's unleashed—and even if he ain't."

＞·＜

Dearest Emma,

I should first tell you—I'm with child. It has brought such a settling feeling and a tenderness from Thomas I had not expected. I'm blessed with a kind husband who brings warmth, comfort, and security to our home. I know he'll love his children and be a good papa. I am hoping in time his love for me will grow as well. I am determined to do everything I can to make that happen.

I'm enjoying work at the mercantile, though I doubt I have your winning way with the customers. I've grown to love Emily. She is a sweetheart, and I'm sure she'll be a good mam. Thomas is well respected by the miners, and I suppose grudgingly so by the mine operators for making them rich. The months here have seemed to fly by. We've not yet had any trouble with the company enforcer, but Thomas remains vigilant. He assures me we will begin our great adventure across the Atlantic very soon. I'm looking forward to it. We will be together, just the two of us . . . well, the three of us.

Thank you for giving me the courage to face my fears, "to boldly enter upon this adventure with confidence and fortitude," as you put it when you pushed me out the door to begin this great undertaking. It seemed such a tall mountain to climb, but life is like that, isn't it? If we take but one step at a time, unmindful of the distance, it's surprising what we can master. I will miss only you, but be warned, I plan on dragging you along every step of the way.

Your sister arm in arm,

Annie

♪ • ♪

Annie read over her letter once more, folded it, then addressed the envelope. Seeing Emily had fallen asleep by the fire, Annie turned down the lantern. She wrapped a shawl around her shoulders, picked up her blanket from the cozy little corner in the shanty she and Tom had made for themselves, and stepped outside into the night. She found her husband sitting on the bench, gazing out at the rising moon.

"It's cold out here," Annie said as she joined her husband on the bench. "I thought you could use this blanket."

They pulled it up around the both of them as they snuggled in together.

"How's Emily?" Tom asked.

"Counting the days. She misses her husband desperately. With their baby coming and all, it's been very hard."

"Well . . . Will and her brother finish their stint of hard labor next week." It grated on Thomas just to acknowledge the travesty of justice. "I'm sure Will is more than excited to get home to her."

"Yes, I suppose he is." Annie couldn't stop thinking of how Emily must feel. The dangerous working conditions at the Oaks had not changed. "What will happen with Will and Mica when they return? The uproar they caused over safety at the mine seems to be a drumbeat these operators and landlords no longer want to hear from your family."

"Yes, and I'm afraid conditions are no better now. Maybe worse, but I promised Tutor I'd stay out of it. I spoke to him about work for Will and Mica. He's committed to give them their jobs back if they want them."

She threaded her arm into his and leaned her head on his shoulder. "I'm sure they'll appreciate your help, Thomas."

He blew out a breath. "Can you believe it's November already?" He looked down at his wife. "Our first baby well on the way, and we've saved enough money to take the biggest step of our lives." He took her hand in his. "How are you feeling?"

"Like a lion." She smiled up at him. "Starting to feel like a mam. I'm a bit anxious, but then as Edie says, 'Anxious is just the flip side of excited.' I'm stickin' with the excited side."

"Good!"

She kissed him on his cheek.

With his arm around Annie, Tom gazed into the night sky. "Think about it, that same moon has just finished shining on our future home in America."

"True enough—it's morning in America," Annie reaffirmed. "I'm glad we promised your mam we'd stay until after Christmas, but I can't wait for us to leave this place. I'm afraid for you." She squeezed his arm.

"Production in the mine is going well, but the tension is thicker than butter. Ole Boo is like a mad dog pullin' on his chain. I see him wandering around the mine more often now, just straining at that leash."

Annie's brows knit together in concern. "Oh, Thomas, I'm ready to go as soon as you tell me. I'll miss your family, of course, and Emily. They've made me feel welcome at the mercantile, but I'm ready."

Tom nodded, still looking out into the night at the moon, his thoughts far away. "I've been reading up on land in California. Rugged, open land for the taking, where we can hack out a life from the wilderness and raise our family any way we choose. Can you imagine, authors of our own destiny?"

"It sounds wonderfully perfect." Annie looked over at her husband with a twinkle in her eye. "But then I suppose perfect will have to do!"

"Perfect it is then. After Christmas we're outta here. On our way to our new home, in a new country, in an untamed land. Likely requiring us 'to keep our wits about us,' as Tigger used to say."

Chapter 26

≻ · ≺

IT WAS TWO WEEKS BEFORE Christmas, and the day began not unlike most winter days in the village of Hoyle Mill. A foggy mist drifted through the valley and gripped the bitterly cold and wet morning.

Enthusiasm for Christmas this year was running high with both Emily and Annie pregnant with their firsts. Christmas was to be held at Martha's, with Edie's five little ones already beside themselves with excitement. Cousin Will and Mica, recently released from prison, were back to work at the Oaks. Will had been putting in overtime to pay off bills accumulated over the past four months while he'd been away working on a chain gang as penance for his indiscretion of confronting Mr. Thomas Dymond and Lord Fitzwilliam over working conditions at the Oaks.

"Will's so excited," Emily told Annie as they prepared breakfast together. "I know he's plannin' somethin' special for me and the baby. But despite my pryin', he won't divulge a thing. Never seen anyone so excited 'bout Christmas afore." Emily looked down at the bangers and mash cooking on the hearth. "I was sore afraid when they took him away from me. I love him so much—it's pathetic, really."

Annie smiled. They were so darn sweet together. "And how is your plan for Will's gift coming along?"

Emily grinned, but before she could respond, Will opened the front door, coming in from the cold. Rubbing his hands together in front of the hearth, he nudged his wife tenderly.

"What are you two plotting?" Will asked.

"It's just for us girls to know," Emily teased.

Will smiled at the two co-conspirators standing at the sink.

"Annie, if I ain't told ya afore, I so appreciate you and Tommy agreeing to stay here with Emily durin' our awful time." He looked over at his wife. "Emily's grown to love ya like her own sister, and with our little one comin' . . . Well, both you and Tom are very special to us."

Will reached over and grabbed Emily by the waist and pulled her close, putting his arms around her and kissing her affectionately on the back of her neck. Annie looked at them both with love. Since Will had returned home, they'd both been like newlyweds, unable to keep their hands off each other. Annie wiped at the moisture in her eyes.

"Annie, you're the best thing that's ever happened to Emily," Will declared.

"Well, maybe not the *best* thing," Emily interrupted, giving her husband her most demure smile. "But ya sure been a godsend. Not sure how I coulda managed without you and Thomas." Emily's eyes gleamed as she looked over at Annie.

Annie's heart warmed. "Of course, we feel the same about you both. It's been a blessing to be able to stay here with Emily these past few months." Annie looked at them with a warmth in her eyes. "Thank you, Will. I can't wait to see your face when Emily gives you your Christmas gift."

"What?"

Emily laughed. "Won't you be surprised," she gave her husband a mischievous giggle.

"What are these two plotting, Will?" Tom asked as he backed in through the door, carrying two buckets of water from the well behind the row of shanties. "Never mind, it's probably more than I want to know."

Both Will and Tom left before dawn that morning, sharing a final word with Emily and Annie over tea and crumpets, bangers and mash, and blood sausage. Both kissed their wives, gave each a warm hug goodbye, and began the short walk from Hoyle Mill to the Oaks Colliery together.

On the darkened path, they were joined by the other collies, chatting, laughing, and sharing jokes with each other in casual conversation. Passing through the gates and into the Oaks yard, Tom headed into the office to cover for Tutor, who was away at a meeting with Mr. Dymond at Woodhouse Mansion. And Will headed toward the cage, where he would be going down into the mine, just as they did every morning.

One by one, each cage filled. Will was lowered down the telescoping shaft. It stopped periodically to deposit miners at various levels, leading into the vast web of interconnecting shafts, tunnels, and chambers, until Will, Mica, and the remaining miners reached the bottom at a thousand feet below ground, where the cage door was pulled open for the final time. They stepped onto the rough-hewn floor of the landing in the faint light. This brotherhood shared their final morning pleasantries, smiling, laughing, and even singing Christmas songs as they each lit their oil lamps and headed off into the various corners of the mine.

⊱ • ⊰

The first trembling upheaval arrived late that clear blue morning, the cruel whip of nature bringing a sudden *crack* of doom to the quiet tenor of innocence as it fought its way to the surface.

In his office, Tom jerked his attention to the window to see a monstrous skyrocket of flames and smoke shoot up in every direction from the main entrance to the mine. Bricks disintegrated into powder. Great ascending clouds of chalk, coal, rock, earth, and debris spouted up into the sky. In an instant, the hundred-year-old cage support beams went up like matchsticks in a blaze. It seemed the entire surface of the earth had been torn off, burying four hundred miners in the pit.

Tom rushed out of his office into the wings of the hot breeze cutting through the cold morning. There was the sound of pandemonium. The smell of smoke and burning flesh. The taste of coal dust. Men yelling. The crackle of burning timbers. Moans of pain and screams of agony.

The massive tree in the yard had been uprooted; fat tendrils of roots rose up from the ground looking like an upside-down tree. The fires pooled and strutted, flowing from structure to tree as smoke chased ash into the sky. The appetite for oxygen was such that leaves and branches were sucked into the flames and flashed their disappearance in an instant.

Adrenaline pumping, Tom leaped into the chaos, ignoring the smoke and flames, seemingly contemptuous of the danger. The main entrance to the mine that was supposed to lay before him was no more. The entire yard was unrecognizable. He had been through mining explosions before, but nothing like this. It looked like a war zone after a daylong cannon raid.

Driven to a rush of heightened awareness, Tom seemed to have a clear vision of just what needed to be done. He knew it was important not to lose his head—to take immediate but thoughtful, deliberate action, carefully planning the dangerous rescue of the men still down in the pit.

His right hand cupped over his eyes to see through the heavy smoke and debris drifting back to earth, Tom held a kerchief to his nose. He searched through the smoky haze to commandeer rescuers from the miners, who were scurrying in all directions in terror.

"You men!" Tom called out. "Come with me!"

The men jumped at his orders, thankful to have someone take charge and tell them where to go and what to do.

✦ • ✦

There were few Christmas shoppers that quiet morning at the Hoyle Mill Mercantile. Annie, who was well into the second trimester of her pregnancy, was determined to work through the nausea without complaint as she stocked shelves for a sympathetic store owner, Mrs. Gray.

Annie was too busy to notice that the mockingbirds had stopped singing and a hush had come over the village square, covered in a thin layer of snow from the night before. When the first weak tremor rumbled up through the mercantile's wooden floorboards, she froze at the muffled vibration and creaking timbers.

Then a deafening *crack* ripped the air. A seismic wave rolled through the village. A frightening whiplash nearly knocked Annie off her feet. The shop's metal boiler tolled like a bell, and the nails in the timbers shrieked and sighed. Dishes, pans, and boxes cascaded from their shelves to the floor in a tumult of confusion.

Instinctively, Annie put her hands over her belly to protect her little one. Her heart pounded in her chest. Her hands trembled, and her ears resounded with the horrifying blast. Leaning against the shelving, she tried to maintain her balance on wobbly legs. She looked through the cracked glass of the window and saw the street buckling, rails splitting, pieces falling off buildings.

Then turning toward the colliery, a quarter mile down the gentle slope to the valley floor, she panicked at the thought of her husband in the midst of it all. The mine was exploding before her eyes.

"Oh, my God!" she whispered, holding her hand over her mouth. But order took over her mind.

Annie took in a deep breath and leapt into action. She grabbed bandages, salves, morphia, medicine, whatever supplies she could pull from the shelves. She threw them into a box, then into the back of the Hoyle Mill Mercantile delivery wagon, and jumped in herself.

Annie snapped the reins. "Let's go, Bonnie!"

The horse leaped forward, and the wagon sped down the cobblestone lane, stopping only long enough to pick up young Doctor Blackburn, who had recently taken over as the only doctor in Hoyle Mill. He was not much older than she, with his rosy, red cheeks.

The wagon clattered wildly down the cobblestone streets toward the mine, bouncing and sliding around corners. The morning sun smoldered through the smoke drifting across the sky. Giant grey mushroom clouds rose above the valley floor, casting darkened shadows over the Oaks Colliery and the village of Hoyle Mill. The coal dust and bloody debris drifted back down to earth like strange, black snow, silky soft, bloody. Annie wiped at her face with the back of her sleeve to keep the grit out of her eyes, while trying to control the sliding wagon racing down the slag road. The thickness of the air made it difficult to breathe.

Passing through the Oaks gate, they rushed into the yard and slid to a stop around the corner from the canteen. Lying before her was a landscape of unrecognizable terrain: distorted steel, collapsed buildings, splintered and burning timber strewn everywhere, people and animals torn into bits. How could the morning, filled with the sweetness of Christmastime, fresh white snow on the ground, now be so tainted with horror?

Men staggered toward them, away from the blazing fires of burning timbers where the main entrance to the colliery had once been. They reeled in shock and confusion, some calling out in agony, some held up by other men, some saying nothing at all. Wherever she looked, torn and broken bodies and injured animals were shaking in the grip of death, while the uninjured tried in vain to comfort the dying. One miner walked toward her wagon, staring dull-eyed at her without a word. It took her a moment to register the man was near naked, his skin burnt grey.

"Lord God," she whispered as she looked over the awful scene. "Please, don't forsake us." But amid all the horror and chaos, God didn't seem to be listening.

Annie's teeth chattered in the bitter cold, but so great was the horror, she was not aware of her shivering. Smoke and ash enshrouded everything in sight. If Annie had been able to conquer her shaking knees, she might have jumped from the wagon and run back up the dark road, back to the refuge of the Hoyle Mill Mercantile. All seemed a fiery hell.

And . . . there was no Thomas in sight.

She shrank closer to Doctor Blackburn sitting beside her, gripping his arm in fingers that trembled, with agony ripping at her heart. She was desperate to know what had happened to the one person she loved more than life itself. She could not survive without him. But she couldn't think about that now. She looked to Doctor Blackburn for comforting words, for consolation. In the unholy crimson glow of the inferno, his profile sat as silent, still as a marble statue.

"God forbid!" Annie stared at the doctor shocked into silence. "You are more scared than I am."

At her voice, the doctor turned to her with wide, vacant eyes sunken into his pallid, blood-drained face. Amid the cold hell of agonizing screams and desperate groans from men and boys, the good doctor was frozen into inaction.

In the calmest, most decisive voice she could muster, Annie called out to him, "Doctor Blackburn! You must do something!" She was trembling, but there was no weakness showing in her face.

He stared at the strange carnage, the raging chaos, overwhelmed by the immensity of the tragedy. Annie grabbed his arm and shook him, trying to drag him back into the moment.

"Doctor Blackburn, what do you want me to do?" she shouted.

Finally, the good doctor began to regain his senses, and shifted his faculties into action.

"Make room in the canteen for the injured," he muttered in a fractured voice. Then in a clearer voice, he cried out, "We will need every flat surface you can find to make the canteen into a triage. I hope to God . . ." he trailed off.

Annie jumped from the wagon and headed in the direction of the canteen. Turning the corner, she faltered in her tracks, shrank back, and clamped a hand over her mouth. Broken bodies sat in the yard, leaning against the wall outside the canteen, moaning, lying on the frozen dirt, crying for their mams in the brutal morning cold. Shoulder to shoulder, bloody, torn, and burnt. Some were writhing in pain. Others lay stiff and still. The smell of perspiration, blood, burnt flesh, corn whiskey, and excrement confronted her in a wave that strained her senses. Men hurried among the dead and dying, bringing in more bodies hauled up from the mine.

Sick at the sight, Annie swallowed hard, raised her skirts, and pushed through the men lying on the ground toward a knot of dazed young men standing idly by the door to the canteen.

"Lady, water! Please, lady. Water! For the love of God, I need water!"

Voices croaked as she picked her way through the dying men and boys. Wincing, she pulled her skirts from their clutching hands. Twisting away, she stepped on one of the miners, and quickly apologized, only to realize when she looked down that the young boy was dead. She stepped over him, over another miner with dull eyes, hands clutched at his belly, dried and frozen blood gluing his clothes to his wounds, his beard frozen stiff with blood.

She tried her best not to be hysterical, to control her fears and direct her attention toward the frightened boys standing by the canteen door. If her heart would stop drumming, maybe she could think.

"Help me get these injured men into the canteen so the doctor can see to them," Annie shouted out in a firm, commanding voice to the dazed group of boys.

"The door's locked, ma'am," a boy answered.

"Well, bust it open, then!"

They did as commanded, nearly ripping the door from its hinges.

"Come with me. Clear the tables and all the counters," Annie barked out as she rushed through the doorway of the canteen.

Seeing a child in the yard just outside the door, she swept him up into her arms, brought him in, and laid him down on the counter. She began pushing everything from the tables to the floor. Dishes, utensils, everything came crashing down. In moments, others were following her lead, relieved to have something to do. Still others joined her in wiping down the flat surfaces to prepare for the injured.

"You boys, get those miners in here!"

They dragged the mangled bodies from the yard into the canteen.

"Okay. On the count of three . . ." Annie ordered those who could help lift the burnt and broken bodies to anywhere a flat surface had been cleared.

While Doctor Blackburn attended the injured, Annie worked to organize the frenzy, frantically directing traffic in the impromptu medical center.

"Leave the critical at this end of the canteen," she shouted, "the serious in the middle, and minor injuries at the far end. We don't have room for the dead, take them back outside—place them in the yard as orderly as you can."

More help trickled into the canteen triage as the morning wore on. No one but the doctor had medical training. Some stood transfixed by the crimson horror, unsure how to help. Despite everyone's best efforts, the shock of the carnage was just too overwhelming. Some just stood trembling and helpless. Others, seeing a young boy brought in with half his face blown away, lapsed into silence. Still others saw the shattered bodies all around them and just stood crying, their hands shaking so badly they had to abandon the work and walk out of the canteen. But a few stayed, struggled through the shock, regained their faculties, then like Annie, did what must be done as best they could with the burnt and torn flesh, dismembered bodies, and blood everywhere.

Doctor Blackburn rushed from table to table, barking out orders. Annie did her best to comply while also organizing the chaos, but she was no doctor—she really had no medical experience at all.

"Hurry up, Annie, I need those supplies," the doctor yelled.

Flustered, she rushed to open a package of instruments and medical supplies, but her anxious fingers fumbled under the harried stress and frustration, spilling everything out onto the floor. She was hyperventilating. Taking a deep, shaking breath to calm her nerves and steady her buckling knees, she swooped up the instruments from the floor, piled them on a canteen serving tray, and delivered them to the doctor. He was amputating arms and legs, clamping arteries, and treating the severely burned. Annie rushed furious to those awaiting her instruction as they carried miners through the door in desperate need.

Men were dying, everywhere screaming in pain, begging for help. But there was little help, and the supplies were dwindling fast.

"Annie, please, I need that tourniquet now. Can you find me a tourniquet?" the doctor demanded in the chaos.

"Doctor, we've used them all." Annie looked around the room at all the amputated limbs.

"We'll, improvise. I need it now."

She saw a belt on a dead miner lying face down on the counter. It wasn't going to do him any good now. She ran to the counter, clumsily grabbed at him, turned him over to unbuckle the belt, and looked into his still-open eyes.

This was no stranger. Annie's legs went limp and she found herself on her knees in front of the table. She stretched out her hand to close the eyes of Billy Abbott, the sweet, intelligent young man who shyly stuttered his way into Emma's heart at Barnsley Dry Goods.

"Annie, where is that tourniquet?" the doctor shouted, his shirt and trousers as red as a butcher's. Even his day-old beard was matted with blood. His was the face of a man drunk with fatigue, impatient with rage, and burning with sorrow.

Annie unfastened the belt and ran it to the doctor, then turned to one of the men. Pointing to Billy Abbott, she said, "Take this this young man out into the yard. He's dead."

Annie knelt over a boy whose body was blistered, his hands and arms burned grey, his hair gone. She was too afraid to even touch him for fear of doing more harm than good. Gently cradling him on her lap, she tried to apply salve, but charred skin peeled off in her hand. He stared up at her in shock, the nerves in his skin so badly damaged he was anesthetized to the pain. His eyes pleaded for help, to save him. But there was little she could do.

Please, God, don't let him see me cry, Annie prayed in her pounding heart. *Help me to be strong for this boy. He's so young. I can't let him see this useless woman with tears in her eyes, looking down on him in the last moments of this earthly life.*

"Doctor Blackburn," she called out frantic across the room. "Please, get over here and bring the morphia. Please hurry, Doctor."

But Doctor Blackburn was literally holding in his hands the life of a young boy with a serious gash in his neck. The doctor had been trying to clean and bandage the wound when, all of a sudden, the boy's carotid artery ruptured. Bright red arterial blood spurted all over the doctor's smock. Reacting on instinct, the doctor shoved his right hand into the hot, slick wound, found the sputtering artery, and plugged the hole with his finger. His eyes were locked onto the blue-green, panic-stricken eyes of the dying boy.

Without words, the boy surrendered his trust into the hands of this doctor, his eyes pleading for him to help. But Doctor Blackburn was not God; he could do nothing but stand over him, feeling the warm slickness on his hand and the fading pulse of life through his fingertips. The boy's hopeful, innocent eyes were fixed in unrelenting fear onto the doctor's.

Doctor Blackburn began to hyperventilate, knowing there was nothing he could do but prolong the inevitable. His anxiety served to add more fear to the boy's pleading expression.

Annie glanced up to see the doctor's face contorted with hate and rage. But it was not directed at her; it was at a world where such terrible things could happen.

"There is nothing I can do, Annie. He'll die. I have no more bandages, no magic salves, no more chloroform. No more morphia." Then in further frustration he added, "Oh, God, to have some morphia. Just a little for the worst ones. Just a little chloroform. Damn these aristocrats who neglected their duties to these men. He was just a little boy, for God's sake," the doctor said, looking down at the child who was now gone.

Tears of fright came to Annie's face as she held the burnt body of the other boy in her arms. Her clothes, soaked with perspiration, were sticking to him. She had been thrust into a nightmare where there was no place to run for safety.

"It'll be all right. God is with you," she repeated to him over and again, offering a forced smile through her tears.

And there he died in her arms.

"I didn't know what to do," she cried out in helpless anguish. "I'm not a doctor! I'm just a nobody—a stupid woman who doesn't know anything. Oh, God, please forgive me!"

Amid all the chaos, a compassionate miner put his hand on her shoulder in a brief moment of comfort. "God will bless you. He sees what you're trying to do," he whispered.

Quiet and careful, the miner lifted the dead boy's body from Annie's arms and carried him out into the yard. Her empty arms still cradling where the missing boy had lain, Annie struggled to regain her composure and move on to the next one in need.

Chapter 27

N EWS OF THE EXPLOSION AT the Oaks Colliery flashed through the close-knit mining community of South Yorkshire. Tom was relieved when Parkin Jeffcock arrived, one of the most respected mining engineers in all of England. After receiving Tom's telegram, Parkin had dropped everything and rushed to help. He brought with him all twelve of the engineering technicians from the Woodhouse Engineering Company, arriving very late that first night ready to go right to work. Tom, having been the Silkstone & Elsecar Coalowners Company engineering manager in his past life, knew the men and considered himself a close friend of Parkin.

Parkin was tall, handsome, with wide shoulders had muscles too heavy to be the aristocrat he once was. He had been thrown out of one of the most prestigious universities in England and one of the most influential families in all of South Yorkshire without a cent. For what reason, no one was exactly sure, and too afraid to ask, though some said it involved an unpleasant scandal with a respectable young lady of nobility. Nevertheless, he was able to attain an engineering degree from Putney College. Known generally for his bravery in the face of many mining disasters, Parkin's closest friends, including Thomas, frequently witnessed his criticism of arrogant mine operators and their appalling lack of safety standards.

"Parkin Jeffcock," Tom acknowledged, reaching out to shake the hand of his longtime friend and colleague.

"It's nice to see you again, Tom." Parkin shook Tom's hand. "When we got your message we all came straight away."

"Thank you," Tom said, nodding to acknowledge the others as well. "Tewart, William, gentlemen. Thank you!"

"Just tell us what you need."

"If you'll follow me," Tom instructed the twelve men, who didn't hesitate to fall in step beside him. "I'll take you to Shaft #3, the only one open right now," Tom shared as they walked briskly across the yard. "Hopefully you can do something to get more ventilation down that shaft, maybe bleed some of it over to the main."

"I understand you've gotten yourself married again," Parkin said as Tom led them across the yard in long, purposeful strides.

"Yes, and we have a little one on the way."

"Congratulations."

They arrived at the entrance to Shaft #3. "Gentlemen, give Mr. Parkin Jeffcock here everything he needs," Tom ordered. "He's here to take charge of the business of saving lives."

"Yes sir, Mr. Wright."

"Now if you'll excuse me," Tom turned from these newcomers.

"Won't you join us?" Parkin asked.

"Maybe later, but right now I've got a bit of business with Skeets and his crew at Shaft #2."

"Thank you, Tom. Good luck!" Parkin tipped his hat. And with that Tom was gone.

"Tell me what you've got here, Skeets?" Tom asked as he joined them.

"Well, Mr. Wright, we're hearing somethin' down that shaft."

Tom leaned over the shaft, and with the rest of the men, listened carefully. He could hear the faint ring of what sounded like a bell.

"Skeets, can you hand me that flask in your back pocket?"

Skeets pulled the flask of brandy from his pocket, took a swig, then handed it over.

"Hand me a bucket, and let's lash as much rope together as we can," Tom directed, then placed the flask in the bucket. "All right, boys! Grab the head gear crank, lower these libations down the seven hundred feet, and let's just see what happens."

When they pulled the bucket back up, the flask was empty. A note that read, *Wanna go home.*

"Sorry, Skeets. Looks like someone down there has a taste for good brandy," Tom smiled with a bit of optimism. "All right, men. Let's beef up the winding gear and lower that contraption down with a rescue party."

There was a nervous excitement as the men rushed to put together an improvised cage and winding gear. While these miners worked, the voices of trapped miners wafted up from the fathoms below singing, and it wasn't long before they joined in the singing. *"Abide with me; fast falls the eventide; the darkness deepens; Lord, with me abide . . ."*

It was after midnight before the miners had completed their makeshift contraption to lower a cage down the shaft over the winding gear.

A half dozen miners gingerly lowered Tom and Skeets one at a time in the improvised coal bucket into the depths of the mine. Because the pumps weren't working, water poured in on them, soaking Tom the whole way down.

When Tom reached the bottom, he found Skeets helping Samuel Brown, soaked and freezing. Mr. Brown had crawled over the corpses of his friends to get to the shaft, dragging others with him and sharing what little food he had. To keep their spirits up, all who were able sang and rang the bell.

"Then came the brandy in a bucket," Sam told his rescuers when they arrived. "Good brandy too. Like manna from heaven."

It was dark, dank, and smoky in the tunnel, smelling of dead bodies and ashen lumber. There were moans from broken men. The splintering tunnel shoring groaned and creaked, precariously balanced on a fulcrum support. It seemed a single fly landing atop the distressed beam could trigger an avalanche of coal, timber, and debris that would bury the lot of them for good.

For hours, Tom and Skeets worked carefully to ensure every last one of the two dozen battered and broken bodies, dead or alive, were returned up the mineshaft to receive medical care or be buried by family.

It was an hour before daybreak by the time all had been safely removed, and an exhausted Tom hauled back up to ground zero. The scene had greatly changed in the hours since. Chaos reigned amid a blaze of torches lighting the harsh, cold darkness. The world seemed to have erupted with scared, grieving family members everywhere living a nightmare. They were frantic for word of their loved ones. Tearful women, men, and terrified children walked in circles, in blank-faced trances, looking for husbands, fathers, and sons. As soon as Tom stepped out of the improvised bucket, he was rushed by a crush of humanity. They came at him with a barrage of questions and stories to tell.

In desperation, a young woman grabbed his shirt sleeve. The refection of torchlight washed shadows across her face, dancing in her wide, terrified eyes. "Did ya see my man?" she pleaded, hysterical. "And me boy? He should have been playin' in the sunshine, not diggin' coal for Christmas!"

All looked at Tom with hope in their eyes. Tom was left speech-less. He had no words of encouragement for the desperate faces, filled with fear, racked with terror, begging for hope. For there was none.

"I'm so sorry," he said over and over. "May God have mercy on your soul."

✦ · ✦

Annie, Doctor Blackburn, and an ever-increasing band of Good Samaritans from the surrounding villages had worked all through the dark night, trying to save as many souls as they could in the makeshift infirmary. More doctors, nurses, and supplies arrived from Sheffield as the sun rose in the east. Annie was able to take her first break from the work, and the stench of the dying miners filling her nostrils—her first chance to escape outside into the cold morning air in the hope of finding some word of her husband. She had heard Tom had not been in the mine at the time of the accident. As expected he had been working with the rescue crews, but no one seemed to know where he'd been all that long night.

As dawn broke on the horrific nightmare, her slack, tired frame sagged against the doorjamb. Her faded blue dress hung long and bloodied over the swell of her belly. Bloody bandages dangled loosely from her hand. Sweat mixed with dried blood was smeared across her cheeks and in her unkempt hair. Annie's shoulders drooped, and her arms hung limp by her sides in exhaustion. Her listless expression told of her long night of anguish in the chamber of horrors. Rimmed in deep, crimson circles, her bloodshot eyes reflected cavernous heart-ache, and the remorse of defeat. The color of death, like heavy liquor, left her mind swirling.

Annie felt sick to her stomach as she looked on at the unrav-eled scene of human tragedy outside. In the bitter cold morning, a mournful procession of wandering women and children, devastated families, milled about the yard in a somber quiet, searching the faces of the dead for loved ones. They walked past the bodies of men and boys laid out side by side, head to foot in grisly repose on the frozen yard. Row after row, more than a hundred dead, with hundreds more still missing.

A pocket of tearful family gathered around a broken and burnt body of a son. A mother sat riddled with despair, her legs splayed wide apart, rocking her husband in her arms while her little ones looked on crying, with wide eyes. Next to her, a family gathered around a mother dazed, hopeless, and crying as they prepared to take home what was left of her three little sons.

There would be no more caresses from a husband, kisses from a father, no more "I love you, Mam" from a son, no more childhood memories of a brother. It was unlikely there would even be enough food to eat or money for rent, let alone the means to bury their loved ones in a marked grave. They would be left without someone to keep them warm through the coming cold nights of despair. Annie's heart broke even more.

Almost every man or boy from the surrounding villages over the age of ten had been down in that mine. There was not a single family in Hoyle Mill who had not lost someone or did not still have someone missing.

In the distance, across the yard filled with the bodies of the dead, Annie watched a trembling, young woman walking tearfully through the rows of dead bodies, searching their faces. Blinded by the rising sun, Annie couldn't make her out, but the gait of the very pregnant woman was familiar. Then she stopped, looked to the ground, and turned away from what she saw, now facing in Annie's direction. Annie raised her hand to shade her strained eyes and focused on the woman transfixed in horror.

"Oh, my dear God," Annie murmured, putting her hand over her mouth to stifle the emotion. "It's Emily. . . Oh no. It's Will!"

A very pregnant Emily, shaky and swaying on her feet, stumbled back, then fell to her knees. Her hands came up to her mouth to hold in a scream. The man next to her grabbed and steadied her.

"Noooo!" Emily cried out, rocking back and forth. Violent sobs and heaves racked her body. A flood of tears streamed down her face. Arching her back in pain, she threw her arms up and shouted into the heavens, "How could You let this happen?"

Annie, standing frozen, couldn't pull her eyes from the scene, but what she saw next took her breath away. She blinked twice to confirm her own lying eyes. At length, full recognition came, and pent-up emotion surged to the surface as she threaded her way through the dead bodies to Thomas, who knelt with an inconsolable Emily in his arms. Emily's brother Mica stood beside her with a comforting hand on her shoulder. All three were looking down at the frozen ground where Will lay, a pale, placid expression on his face. There was no disfigurement. A victim of the afterdamp, his entire body was now covered in a fine dust of grey ash and soot.

Annie leaned down and put her arms around Emily.

"They killed my husband, Annie!" Sitting on the ground Emily held her husband on her lap. "Oh, God. He looks as if he's sleeping, like he was this morning in bed next to me."

"Oh, Emily, honey," Annie's grief-stricken voice cracked, but there was no consoling the woman who had become like a sister to her.

How giddy Emily and Will had been about Christmas and their baby that morning. It seemed a lifetime ago.

"He went down in that cursed hole for Christmas money." Emily's voice broke. She sobbed, "To buy me a Christmas present."

Mica and Tom carefully gathered Will in a blanket and placed him in the back of Mica's wagon.

"I'm gonna take my sister and Will to my mam's," Mica said as Annie, still holding on to Emily, helped her into the wagon.

"I'll stop by with your things, Emily." Annie tried to catch Emily's downcast eyes.

"Give us a couple days, Annie," Mica suggested.

There were no more words spoken. Annie put a hand on Thomas's shoulder to comfort him as a tormented Mica left with an anguished sister and her dead husband.

As the wagon drove off, Tom turned and put his arms around Annie. She sobbed against his comforting shoulder. All that she had held inside over the long day and night was released. "So many—so many dead boys, men, and destroyed families. So many more still lost down in that cursed pit. What are we to do? Their wives? Their children? Their mothers?" Annie wept. "What are we to do, Thomas?"

For a long time, Tom and Annie sat on the ground, leaning against the canteen wall holding each other, whispering about what the future held in store for this community after the terror of the day's events. Not a family in Hoyle Mill had been left without a lost loved one.

After several minutes of quietly sitting together, Tom turned, and apologetically asked, "Do you mind . . . I'm so sorry, but I must do something or I'll burst." Anger was rising in him for the uncaring, irresponsible operators of the mine. "It's too much for me to think about. I need to be busy. I need to get to work. Do you mind?" He looked into her eyes for a release, taking another ragged breath in resignation.

"God bless you, my husband. This is a time to be busy with helping hands for those who can still use our help."

"You have done too much already. And in your condition. Will you promise me you'll go back into the village and rest? I'll meet you at the mercantile later."

"I'll be fine."

"Promise me. I can't leave you unless you promise me."

"All right, I promise."

"I'm gonna try to join up with Parkin. They're down in Shaft #3." With tenderness, he kissed his wife, and they parted. "Please, Annie, take some time for yourself and our baby—please?"

She nodded. "Be careful."

Tom offered a strained smile, then turned to join the others.

Numbly, he headed toward Shaft #3. Reaching the entrance, he stretched out his arm to pull the cage toward him when suddenly the earth began to rumble and shudder. The already unstable ledge began to crumble, heavy rock rolling down the unraveling escarpment. A moment later, the entire pit exploded upwards, turning the earth inside out. Coal, rock, and timber flew up in every direction like murderous projectiles.

Annie screamed and dropped to her knees, her hands covering her cheeks as she looked on to where her husband had been standing.

Chapter 28

꘎ • ꘎

ANNIE STOOD GRIPPING THE POST on the porch of the canteen with one arm, the other on her belly. She stared ahead without seeing, waiting. She couldn't bear to be inside with the doctor.

She was jolted out of her stupor when Doctor Blackburn's hand rested gently on her shoulder. Then turning anxiously, she faced him.

"Thomas has quite a gash on his forehead and contusions everywhere. He has severe lacerations to both hands," he told her. "I've sutured up the wounds, and the nurse is putting on bandages. His cuts and bruises should heal just fine with time and proper care to avoid infection."

Annie nodded. She was dazed, solemn and confused.

Doctor Blackburn paused to let her gather her faculties. "You better sit, Annie. This is a lot to take in your condition."

Annie, noticing the doctor's grave tone, feeling woozy with worry, reached for his arm. He held her steady and pulled up a chair.

"I'm afraid Thomas has suffered a very serious head injury. Thank God he put his hands up to soften the blow—it undoubtedly saved his life. But he's unconscious right now and could remain that way for some time to come. We can only hope he comes out of it. We just don't know . . ." The doctor trailed off.

"What?" Annie felt numb. "'Hope he comes out of it?'" Her voice shook as she stepped inside to find Thomas's lifeless body lying across the cold canteen counter under a thin blanket.

She felt like she was looking down from a dream, her mind disjointed from her body. Her heart pounded wildly in her chest, her face pale, drained.

Doctor Blackburn sighed. "I don't know how long he will remain this way. It could be a day. It could be a week. To be perfectly honest with you, he may never wake up at all. I'm so sorry, Annie."

Doctor Blackburn's voice sounded far away as Annie tried to comprehend the weight of his words.

"Right now, the best thing you can do for him is to make him comfortable, change his bandages often, and just wait and see," he offered. "There's not much else we can do. I've packed some salve and clean bandages. Take him to the mercantile. Mrs. Gray can help you through this. If—or, when he does awake, be gentle and patient with him. He might not remember a thing, and if through God's mercy he recovers, he may have a difficult time coping. His body will mend in time, but the rest, well . . . it will take a long time, at best."

"What else can I do for him?" she asked teary-eyed.

The doctor looked into her eyes. "Pray. You can pray with all your heart. Thomas is in God's hands now."

Annie looked down. "I'm not so good at that."

"We'll probably all need to get a little better at it now," the doctor commiserated, looking around the room at all the battered bodies and broken hearts. "Annie, you have been a godsend here. You dragged me out of my stupor, put me to work, and then worked harder than anyone has the right to ask. I'll never forget what you did for me and all of us here. Thank you. God bless the both of you . . . all three of you."

❧ • ❧

Several of the townspeople helped load Thomas onto the mercantile wagon and drove he and Annie into the village. Mrs. Gray gave
Thomas her spare room, so the both of them would have a comfortable place to spend the night. Beyond exhausted, Annie fell asleep in a
chair holding Tom's hand. She stayed this way by his bedside all night
long, even after Mrs. Gray brought in a cot and placed it next to him.

Annie spent the next three days and nights either sitting or sleeping beside Tom while he remained unconscious. In fear for his life,
Annie got little sleep. She sat watching him breathe with her heart
in her throat. Mrs. Gray brought in meals and comforted her as best
she could.

On the morning of the fourth day, while Annie was dozing in the
chair, she heard a raspy, "Hello, Annie."

Slowly, she opened her eyes, lost in half-sleep. He was looking up
at her.

"Thomas?" she quickly sat up straight. "Thomas, you're awake!"

He glanced around the room. "Where are we?"

"Oh, Thomas! I have been so worried," she cried, drinking in the
sound of his voice. She rushed to him, awkwardly wrapping her arms
around his neck, pulling him as close as his condition would allow.
"Are you alright? Your head, how does it feel?" Annie paused. "Do you
remember the explosion?"

"My head hurts."

"Oh, Thomas. I was afraid."

Tom blinked. "What am I doing here?"

"You've been unconscious."

He looked up, confused.

"It's been four days."

"Four . . . ? The others? What about the others?"

Her heart sank. "Thomas, I'm so sorry." Annie leaned forward in the chair, then gently reached out to grab hold of his hand.

"Parkin?"

"The explosion. He's—they're all gone."

"Gone? Gone where?"

Annie just looked back at him, not knowing what to say. "They're dead. They're all dead." Annie tenderly squeezed his hand and leaned down to hold her husband. "I am so very, very sorry, Thomas," she whispered in his ear.

But he said nothing.

She pulled back, awkwardly waiting for him to speak. His expression was filled with confusion, staring back at her, not comprehending. "I don't understand."

"Your friends were still down in the mine when the second blast hit. The eruption . . . it was like nothing any one had ever seen before. The whole earth shook, the smoke, the flames . . . from everywhere. We thought the world had come to an end." Annie ran on nervously, afraid if she stopped and looked into his eyes, she'd start sobbing. "Tutor tried to rescue them, but then a third explosion ignited the gases and destroyed all hope of further rescue. There's nothing left but a pile of rock and charcoaled debris. All three entrances into that cursed hole were sealed off. Hundreds of men and boys left inside." She bit her lower lip, grabbed Tom's hand, and stared into his vacant eyes, shattered red, rimmed in scarlet. She waited for his reaction.

He lay silent for a long time. But no more words came.

⤜ • ⤛

Thomas spent more than a month recovering in bed at the mercantile. Mrs. Gray had insisted after Emily had moved back in with her parents. When Tom was well enough, Annie collected their things, and with Joey's help, they made the long drive back to his family home. A portion of the family's barn had been prepared into comfortable living quarters for Tom's long road to recovery. His lacerations, cuts, and bruises were on the mend, and even his head injuries were miraculously recovering. But the trauma had left him with a clouded mind. He was confused. Frustrated much of the time, but unwilling to share his fragile emotions. Annie, Martha, and Edie flitted around him, trying their best to make him comfortable. Still, it seemed neither man nor woman could follow him into the nightmares where his mind traveled.

"How are you feeling?" Annie asked one cold March morning.

Tom forced a smile but was reluctant to discuss his disoriented feelings. He didn't want to talk of the depression that gripped his heart for fear of what it might do to Annie.

When Tom didn't speak, worry rose in her. When he did, his disconnected thoughts scared her. Tenderly, she doted on him.

"I'm just concerned about you," she said when he got frustrated. "Edie is making you some warm soup. Maybe I could read to you. Would you like that?"

"I don't know. It's awfully hard to think of anything but all those men. And their families—what's happened to their families, Annie?"

Annie sighed. "We're doing what we can to help. Please, let's get you better first, before . . ." Annie paused. "We're holding off visitors until you're feeling a bit better, but several have asked after you."

"I can't help but feel guilty. I should have done more to prevent . . ." he trailed off.

"Tutor stopped by to ask after you a couple times, and some of the miners and their families too. Maybe when you're a little stronger you can see them?"

"The company?"

"We've heard nothing."

"Sorry, Annie." Tom looked at her plaintively. "This is a hell of a way for a husband to be taking care of his wife and child. I'm not much fun, sleepin' right through Christmas, and New Year's too." Tom sat in bed, clearly frustrated.

"There will be other Christmases." Annie smiled down at him, but he looked off into the distance, staring vacantly.

"My cousin Will is dead—did you hear? We were close."

"Yes, I know. I'm so sorry."

Tom prattled on. "Will looked asleep lying there on the frozen courtyard. There wasn't a mark on him. They found him in the arms of another miner," he muttered. "But then I suppose you know all that, don't you?"

"It's gonna be all right, Thomas. You were a real comfort to Emily," she responded. "Did I tell you I went to help Emily with the delivery of her new baby? What a beautiful little thing she is. Surrounded by family. Emily seems well enough, all things considered. She's gonna be a great mam. She would love to come see you when you're feeling better."

Tom's head was bowed down. "Parkin's dead." Tom looked up at her bewildered. "Tewart and William are dead. All twelve of them are dead because I asked them to come." He pinched the bridge of his nose as if in pain. "They're all gone. None of 'em ever coming home to their families again."

"Oh, Thomas." Annie took his hand and held it in hers, trying her hardest to be supportive and positive, but she was afraid for him.

"I feel so guilty—what am I to say to their families? They trusted me, and I took them to the shaft that killed them . . ." Tom trailed off, emotion brimming to the surface. "If not for you and the baby . . . it's so much harder knowing I'm here, and they're gone forever."

"It's not your fault. Everyone has said so. You and the others tried to save those miners."

Tom just stared out the window into the yard. He'd known for years the Oaks was a death trap. He warned the company—several times, in fact. But he should've done more, he and Parkin.

"Thomas, when I thought I was going to lose you, my heart broke into a thousand pieces. I couldn't stop bawling like a baby. But you're safe now. We can try to help those rescuers' families, but you must get better first."

He sat quiet in his bed for a long time as Annie held his hand.

Chapter 29

＞ • ＜

ANNIE SAT IN THE FARMHOUSE window seat, arms wrapped around her legs, her chin resting on her knees. The soft light of the summer morning streamed in silent to warm her as she watched her nieces and nephews at play in the yard. After days of numbness, she had succumbed to the growing roots of melancholy that pressed against her chest.

Annie's baby had come early on a cold April night. It might have been the stress from the tragedy in Hoyle Mill. Or maybe it was the burden of long hours caring for her husband. Or the nights of holding the hands of the mothers who would never again see their husbands, the long days comforting families left without papas or means of support. Whatever the reason, by noon the day after his delivery, Baby Henry had developed a fever.

"He just lies in my arms staring at me wide eyed, whimpering," Annie cried to Martha. "I can do nothing to make him suckle!" She was in tears after a long day of trying to coax Baby Henry into nursing.

For five days and five long nights, Annie had held him. She was exhausted. Martha and Edie tried to help in every way possible. Even an exhausted Thomas tried feebly to lighten the load where he could, but no one could forestall the inevitable.

"I know he's hungry, but what am I to do? I cannot get him to take my milk," Annie lamented. "He can't seem to find the strength."

"I'm so sorry, honey," Martha tried, helplessly, to comfort her.

In the end her baby was too exhausted to fight any longer. As life slipped from Annie's babe in arms, so went the light of joy from the young mother's heart.

Tom had slid in beside Annie and held them both in his arms. But it was difficult to push aside the feeling that their world was collapsing around them. Both Annie and Tom were confused, distraught, and felt helpless to do anything.

"I pass you now to your Father in heaven," Annie whispered into Henry's little ear, too exhausted to cry anymore. Struggling to regain her composure she added, "I wish He could hold you in heaven for me."

When Martha took the baby away, an unfathomable grief came in great rushes like waves of suffocating water passing over her. Then came great sobs and ragged heaves. The pain clung inside her like a cocklebur that dug deeper when she tried to pull it free. So, Annie pulled and the pain dug deep into her heart, where she was convinced it would remain forever.

"How could I not save my baby?" Annie repeated delirious as she lay heartsick in bed, exhausted, disillusioned. "What kind of mother am I? I couldn't even feed my baby."

"There was nothing more you could do. No one could," Martha tried to comfort her daughter-in-law. "This is one of the trials of motherhood. The love comes regardless. Your heart is never too broken to love your little ones, even when their lives are not yours to keep."

"It just . . . hurts so much." She placed hands on her chest, "Deep inside me."

Martha was silent for some time. "Of course, it does—it's motherhood." She sighed. "But what can we mothers do? Who can measure a mother's grief at the loss of her child?"

Martha leaned down and put her arms around Annie, who returned the embrace.

From the moment Thomas had brought Annie home, Martha had become Annie's mother in every way that mattered. Even if they didn't talk of their love or share their feelings openly, the bond was always there. They were both strong women in their own way. Women who worked hard—doers who seldom spoke of heartache or sacrifice.

Annie struggled in quiet resignation, afraid of breaking the dam that held back her overwhelming grief and unleashing the despair from her shattered soul on her struggling husband. She did not want to lay her hurt on his shoulders and derail his difficult recovery.

Thomas, her stalwart, stoic coal miner, tried to be strong for her in the only way he knew how. He turned his own emotions inside and in his own way addressed Annie's need to be comforted by leaving her to herself. She knew his own grief screamed out for help. But she couldn't shoulder it for him.

Little by little, Annie and Tom resumed their daily life together. Until finally, Annie had to pour out her heart in a letter to Emma.

Dearest Emma,

My Henry died, as babies do, gently and without complaint. Maybe it's because they have spent such little time with us that they seem to hold to life so weakly. I wonder if it's the memory of heaven still living so fresh within them, they don't seem to fear death as we do.

They told me on that last night when our Henry struggled for life, "Watching your child die is the longest nightmare of your life." They were right. But now the nights seem endless, agonizing, and only lead to torturous mornings.

I can't get the thought out of my mind. I held Henry and watched him take his final breath in this life. I can't even talk to my husband about it, for fear of adding to his already heavy burdens of his plaguing injuries. I know he, too, is hurting from Henry's death. He struggles to push back on his bouts of depression for my sake, but we don't talk to each other about it. I know he is trying to be strong for me. I can't bear to tell him it's not working.

I have heard that after you lose an arm, on occasion you still feel pain in your missing limb. At night, I wake up not only with longing, but with an awful ache to hold my little one. It is hard to imagine how it's possible, but my empty arms ache for the loss. I find myself unconsciously rubbing them to sooth the pain, but the ache remains, nonetheless.

Some days I just want to spend the whole day curled up under the covers, but of course that's not possible. I try to keep my tears to myself for Thomas's sake, but with my emotions so close to the surface, I sometimes break down for no apparent reason. I know he knows but is afraid to ask me about it.

I don't think I will ever be able to put aside thoughts of my Henry alone in the cold earth of the potter's field. But I need to be strong for Thomas. He has so much still to deal with.

Please pray for us.

I will love and miss you always,

Annie

Chapter 30

⊱ · ⊰

IT HAD BEEN THE BETTER part of a year since the tragedy at the Oaks Colliery. Tom, nearing the end of his long, difficult convalescence, continued to live with Annie in his family's converted barn. Annie spent her days caring for him, washing and cleaning for the family, helping with the children. If not for Martha and the rest of the family, Annie and Thomas would have been left destitute.

After the Oaks disaster, most of the struggling families found it near impossible to cope without the companionship of their loved ones. In many cases, there was no financial support. Tom and Annie were among the lucky ones. Some widows and orphans resorted to begging to survive. Some, so desperate to provide for their little ones, fell prey to the Candyman, and those like him, who circled like vultures to take advantage of young mothers whose families were suffering impenetrable hardships.

Bishop Walker, who was witness to many of the injustices, took it upon himself to do something about it. He became the self-appointed advocate for those who could not help themselves, paying no regard to a destitute family's particular religious persuasion. He wanted only to do the right thing by these working-class miners and their families who were in need. Quietly, he went about trying to ensure justice was served, enlisting the help of others where he could.

Bishop Walker stopped by the Wrights' home early one afternoon, not long after Baby Henry's funeral.

"Look who's here?" Martha called to Annie, who was just pulling a kettle of tea off the hearth. "Bishop Walker has come to speak with Thomas."

"It's so nice to see you again, Bishop. Would you join us in some tea?" Annie poured a cup of tea for Martha, Bishop Walker, and herself as they all sat down together. "I'm sorry that Thomas isn't here right now."

"We can get together another time. Thank you for the tea, Annie." Pausing, he took a sip.

"Actually, I've been hoping I might have the opportunity to talk to you ever since Henry's funeral. It was such a heartache for all of us. I hope you don't mind, Martha told me how hard it's been on you."

Annie smiled at Martha. "I don't mind. I'm grateful for the support."

Bishop Walker continued. "I wanted to let you know that you are greatly loved by your neighbors. Many have told me they consider you part of their family after all the unselfish acts of kindness you have freely given."

Annie took a sip from her tea and smiled weakly. "I have grown very close to many of them over the past few months. Their friendship is a bright spot in my life."

"Shared suffering can sometimes bind people together, don't you think?" Bishop Walker observed. "And sometimes tragedy makes ties that bind grow even stronger. It's a testament to your giving heart."

"Yes. Shared trials can bring people together," Annie agreed. She looked out the window in solemn contemplation.

"How is Thomas doing with his recovery?" He studied Annie's expression. "It's been a struggle, I'm sure."

Annie nodded. "He's . . . he's making progress with his physical injuries, thank God. Progress is slow, but steady. The mental strain is another thing altogether."

Annie fell silent for a long moment.

"Battling the exhaustive headaches, the dark melancholy—it's difficult for him. Especially with the loss of Will, and so many of his friends, and, of course, there's our Henry. I know Thomas is hurting, but he tries to keep his burdens to himself. . ." Annie trailed off. "He doesn't talk much about it," she sighed. "I suppose he sees his lack of complaint as protecting me."

"I see." Bishop Walker nodded. His brows turned down in a tight frown of sympathy. "That's a heavy burden. The mental anguish must be very difficult—for you both. He's very lucky to have you, Annie."

Her tight-lipped frown softening, Annie quickly raised her brows in acknowledgement. "In some ways, I think he feels personally responsible for the families of the rescuers. And he feels guilty not being able to help relieve some of the strain their care has put on me . . . after Henry's . . ."

Bishop Walker nodded. "How are you feeling?"

"I would be lying if I told you I haven't struggled since the funeral."

"Maybe you and Thomas could come to dinner sometime soon. It would be good to talk, and I'm sure Margaret would love to have you."

Annie's voice quavered at Bishop Walker's kindness. "I will check with Thomas. Thank you." She continued to look away, trying to hide the moisture welling up in her eyes.

Bishop Walker thanked her, then whispered, "Before I leave, do you think I might share a couple thoughts I've pondered on since Henry's funeral?"

Turning back, Annie quietly responded, "I think I'd like that."

Martha gently put her hand over Annie's.

Bishop Walker pulled out his scriptures and began to thumb through them. "Christ tells us, 'I will not leave you comfortless. Let not your heart be troubled, neither let it be afraid.' Then He says, 'Blessed are they that mourn: for they shall be comforted.'"

Annie eyes were hollow. "It sounds so simple. But I don't think I can find peace in the loss of my little one."

Bishop Walker sat silent for a long moment, looking down at his scriptures. Then he replied hoarsely, "Margaret and I lost our first child. I will never forget our Katy."

Annie felt a pang of regret in her heart. "I'm so sorry. That is so difficult." She paused. "I can't imagine I will ever get over the loss of our Henry. It's been quite a struggle to push back on the depression. I don't really know about Thomas . . . He doesn't talk to me about how he feels. Oh, how I wish he would." Annie turned away, wiping the back of her hand across her runny nose and eyes. "I'm sorry."

"It's different for men than for us women," Martha interjected. "Of course, all of us are different, but husbands don't carry our babies for nine months. They don't feel the child growing inside. There is an unbreakable bond between a mother and her child. Henry was truly flesh of your flesh, and bone of your bone. You literally shared absolutely everything with him."

"I suppose that's true."

Bishop Walker nodded. "I would like to think there is a special place in our Father in Heaven's heart for mothers," he added. "It was Jesus's mother who passed through the valley of the shadow of death to bring forth her beautiful baby boy, the Savior of all the world, only to watch her perfect son be crucified. It was Jesus who said to John in the very act of His crucifixion, 'Behold thy mother.' It was as if he'd said it to all mothers."

Annie nodded, so touched she was unable to speak.

"Carrying a child is such a binding experience—bringing you so close to death." Bishop Walker looked into Annie eyes as she slipped into silence, listening carefully to what he was saying. "It is such a traumatic and powerful confluence of emotions. As a mother, you feel and appreciate emotions that no man on earth ever could. Christ tells us mothers have a better understanding of what the Lord feels for each of us, His children. *Bearing, carrying, delivering.* These are such powerful, messianic terms. Words that reflect the burden, the struggle, the fatigue and yes, sometimes the heartrending loss. They are also words most appropriate in describing the mission of our Savior who, at unspeakable cost, bears us up when we have fallen, carries us forward when our strength is gone, and delivers us safely home when safety seems far beyond our reach."

"I've never thought of it like that." Annie swallowed a lump of emotion.

Bishop Walker continued, "I always wonder what my Katy might be doing right now."

"What do you mean?" Annie asked, a grimace wrenching her face. "You wonder what your daughter is doing now?"

"Like the saintly mother of Jesus, you carried a child of God, you bore Henry, then delivered him back to his Father in heaven, complete with his salvation."

"Complete with his salvation?" Annie repeated, lowering her eyebrows in a quizzical frown to question his words. She wiped at her eyes with the back of her hand. "Do you think my son has gone back to his Father in heaven?" Tears were now running down her cheeks, desperately wanting it to be so. "The minister at St. George's says my baby is dead to God because he never received his first communion. He said he was banished to everlasting darkness. You were there.

Henry couldn't even be buried in the church cemetery . . . My beautiful baby boy was buried in the potter's field."

Bishop Walker took Annie's hand in his as she tried to control her emotions. "Dearest Annie, I can testify to you Henry is with God," Bishop Walker whispered. He spoke with such certainty it tore at her heart.

"Oh, my determined, strong, and loving daughter," Martha interjected, "your innocent babe has completed his time on this earth and returned to eternal peace."

Annie looked back, her cheeks puffy and pink, her eyes ringed in scarlet. "With all my heart, I hope you're right."

☙ • ❧

Three weeks later, Tom entered the bishop's home with Annie on his arm.

"I appreciate you inviting us to dinner, Bishop," he nodded cordially to Bishop Walker's wife. "Margaret."

"Well, Margaret and I appreciate you accepting our invitation," Bishop Walker bowed graciously. "You've done so much to help with the families of the miners that we wanted to formally thank you both."

Tom nodded toward Annie. "Annie's the Good Samaritan, not me."

After spending the early evening talking about their work with needy families, Tom changed the subject. "There needs to be an inquest into the Oaks Colliery disaster. And you need to be the miners' advocate."

"Excuse me?" The good bishop was surprised at the bold suggestion. "What makes you think anything can be accomplished by an inquest?"

"Oh, I don't doubt Lord Fitzwilliam, the great Parliamentarian, will rig the outcome of the inquest. But that's why it needs to be done. His arrogance will work to our advantage." Tom's eyes were aflame.

"I don't follow you?"

Tom took a deep breath. "If all the horrific deeds of the mine operators can be played out in the inquest, laid before a sympathetic public through the equally sympathetic national newspapers, the rest of the politicians in London might be forced to do something about our plight. Otherwise, things will never change. Men like Lord Fitzwilliam and these other mine operators will continue to rule over every aspect of our lives, and thugs like Boo Black will continue to get away with rape and murder."

"I see you feel very strongly about this," Bishop Walker said, pondering the prospect. "Why do you think I'm the man to lead this fight?"

"I've known you a long time, Bishop. You are a well-respected member of this community, an eloquent speaker, educated in the same schools as these aristocrats who wield the power in this country. If not you, then who? You have the experience and the tools to do it. I can only hope you have the conviction to risk the fight. Evil is powerless if we but have the courage to stand together and resist."

Bishop Walker was not the kind of man whose lips went dry when under pressure, but what Thomas was proposing was risky business with these powerful men. "Most of these aristocrats either have been or are currently in Parliament, including Lord Fitzwilliam," Bishop Walker said thoughtfully. "But I suppose, as the Bible tells us, pride comes before the fall. Given the opportunity, the unconscionable arrogance of Lord Fitzwilliam and these other mine operators might just sink their own ship."

Tom's eyes lit up as he saw Bishop Walker catching his vision. "And I'll give you everything you need to help sink that ship. In a sham hearing, his handpicked 'Board of Inquiry' will ignore the damning facts and hold themselves harmless, regardless. It's our job to ensure the London newspapers report the story correctly. Public outrage at the atrocities, ill-placed privilege, and manipulation of justice. It may be our only chance to tame the beast and achieve some real change."

Bishop Walker nodded. "How do we ensure the reporters will come prepared to tell the truth? And how do we know the politicians will not ignore their obligation to rebuke the deeds of fellow members of Parliament?"

"The story will sell papers. And politicians would chew up and spit out their own mother for votes," Tom offered dryly. "Parkin Jeff-cock despised all those behind this disaster. If he were alive, he would be the first to admit there are advantages to his relationship with these reprobates and scoundrels who will write the stories in their newspapers. Parkin has left behind many fraternal brothers, who also happen to be well-connected politicians upset with his senseless murder. Moreover, Queen Victoria has proven to be a sympathetic monarch who could make the difference when presented with the facts."

"Even if we're able to get the story out to sympathetic politicians, we'll need some hard facts to expose the company and landlord's actions. How do we get those documented facts?"

Tom sat back in his chair. "Most of what I know is in my head, from years of serving both the company and Lord Fitzwilliam. But there are plenty of damning documents in company files."

Bishop Walker paused. "Can you get ahold of those documents?"

"I can't, but I know someone who can. Of course, he must be convinced."

"And you think he can be convinced?"

"He must be . . . and he will be," Tom declared with conviction.

❧ PART III ❧

The Inquest, Hoyle Mill, 1867

Chapter 31

⤜ • ⤛

AFTER MONTHS OF PRODDING NEGOTIATIONS, an inquest was scheduled to be held in Hoyle Mill. It was to be conducted by the coroner employed by Lord Fitzwilliam. In fact, all but Bishop Walker, representing the interests of the miners' families, were handpicked by His Lordship, including the members of the Board of Inquiry.

On a cold, late December night, one week before the start of the inquest proceedings, Tom and Annie sat in front of the fire after everyone else in his parents' farmhouse had gone to bed.

"I think we should go," Tom whispered.

"I'm sorry, go where?"

"I suppose this might sound like you did indeed marry a crazy hermit, after all. But I think it's time to leave for America to start our new life." He gave Annie a concerned look. "We've overstayed our welcome. There's nothing keeping us here."

Annie sighed. "Nothing but money. That's what's left of our savings—nothing! Of course, I think we did the right thing sharing all we had with these destitute families, but now it's gone. It's expensive to cross the Atlantic, to buy provisions. And you're still not fully recovered. Thank God you're feeling better after a long, hard-fought recovery, but you still have a way to go." Annie paused a long moment and Tom could see she was trying to choose her words carefully for a final objection. "There's one more thing. I haven't known for sure until now . . . but I am with child again. I feel her life inside me. I want to do everything I can to be careful this time."

"Annie, but that's wonderful!"

"I'm afraid, Thomas."

"Oh Annie." Tom sat silent. "You're right, of course, about the money, my recovery." And wow, that's wonderful news about the baby. If I were you, I'd seriously consider committing your husband." He paused. "Still . . . I have the strongest feeling we've put it off long enough. I don't trust the company or our paternal government to decide our future. Both have their own self-interests at heart, and it will never match ours."

Annie's eyebrows forced together into a pronounced frown as she asked, "Is there something you're not telling me?"

"There are rumors, I'm afraid."

Annie flinched. "What do you mean, 'rumors'?" She turned to him with a withering look. "Do I need to drag you to the Tower of London to get you to tell me these things? I'm your wife."

He took a deep breath, then blew it out, "I spoke to Tutor. Of course, the inquest has spiked concern at Woodhouse." He paused, knowing he hadn't been thinking clearly of late. "I thought we were relatively safe out here. Miles from town, out of sight, out of mind. I assumed there were more important things to worry about after this disaster than me, but they went by Will and Emily's shanty looking for me. Tutor told them we left town, which he warned we should. They threatened him, Annie—told him he can't testify at the inquest."

"What does that mean?"

"Tutor is our key witness. It leaves us few options, but . . ."

"But what?" Annie ran her sweaty hands down her apron.

"With Tutor unavailable, Bishop Walker has asked me to take the stand at the inquest."

"Testify? Oh, no, Thomas. You can't do that." Annie shuddered at the implications, her heart pounding.

"I told Bishop I'd talk to you about it. I thought—"

"You thought what? Look at what Boo Black's done to the families of these dead miners," Annie interrupted. She looked at her husband long and hard with concern in her eyes. "I know you want to, Thomas, but I hear the company thugs have beaten men to death for less, and . . ."

"I know you're concerned about me, Annie." Tom sat up to his wife, knee to knee, and took her hand in both of his. "What would I do without you?"

"That's an excellent question!" she looked away frustrated. "Who would defend your crazy ideas?" She turned back to face him. "Oh Thomas, I'm afraid for you."

"I realize it's insanely foolish." He searched her eyes for understanding. "We'd need to leave here immediately after my testimony and on to America as soon as we can thereafter, but mustn't I do this."

"Right—upset these dangerous people, then run dragging your pregnant wife behind you, with no money, in a rusty tin can halfway around the world to some mountainous wilderness."

"Something like that!" He looked at his wife uneasily. "What would you have me do?"

"Oh, Thomas. I suppose you'll never forgive yourself if you don't testify." Taking a deep breath, she looked softly into his eyes. "It scares the hell out of me, but I'll make all the arrangements to leave while you prepare for your testimony with Bishop Walker."

"Thank you, Annie." After a long silence, Tom added, "After we pack up and settle things, I think it's wise for you to go to your family's home in Handsworth until all this is over. I can join you there afterward, and we can—"

"I'm sorry, Thomas, but that I won't do. I'm staying with my husband. As I recall, I signed on for hell or high water."

"Oh, Annie. You may just get both."

Annie nodded. "Fair enough, but we'll take this leap together, and pray to God we can fly!"

>· <

On the following Sunday, Thomas and Annie attended church together at Bishop Walker's request. Tom seldom attended, but Annie had come to look forward to these Sunday morning sacrament meetings. She enjoyed spending solemn time together with Martha, Edie's family, and so many of their new friends. The church service had already started when they slipped in to sit beside Martha. Everyone in the congregation seemed to be paying especially close attention to Bishop Walker on this Sunday, the day before the beginning of the inquest. In the quiet hush of the morning service, a warm feeling coursed through Annie's veins as she held Thomas's hand. She felt an even greater closeness to him in this place.

After the meeting, Bishop Walker came to Tom and Annie and invited them into his office. "I appreciate you letting me take a moment to speak to you both," Bishop Walker earnestly began, gesturing for them to take a seat. "How are you feeling, Tom? Your headaches, how have they been?"

"Much better, thank you," replied Tom as they all sat down.

"I can't tell you enough how much I appreciate your willingness to testify. It's dangerous, I know, but without Tutor, you are the most knowledgeable witness in all of South Yorkshire. Your testimony could make all the difference."

"What is right can never be wrong."

"Thank you, Thomas, both of you."

"We'll just have to be prepared for the consequences. When the rest of the country finds out what these mine operators have been up to, I'm hoping it will be hard for honest men not to force the politicians in London to do something about it."

Bishop Walker nodded in agreement, and then leaned down and pulled an envelope out of his desk drawer. "I've just received a letter that might interest you. I know you've drained all your carefully saved immigration money in helping these families who have no financial support—we all thank you for that." He glanced at them both. "A few weeks ago, I contacted a friend in London. And I just received this response." He opened the letter and went on quickly now. "'The Comstock Investment Company has offered Tom a severance contract for a position as Director of Operations in Virginia City, Nevada Territory, out in the American West.'"

Both were silent as they processed this unexpected offer.

"What's a severance contract, exactly?" Tom asked.

"You'd be indentured at this silver mine until the immigration debt can be paid back, which shouldn't take long with the salary offered. They'll pay expenses, but the one catch is they want you to leave right away."

Speechless, Tom and Annie stared at each other. The Bishop handed them the letter. They both studied it.

"I know where this is!" Annie's eyes widened in excitement.

"Really?" Bishop Walker raised his brows in surprise.

"It's out west in the Sierra Nevada mountains, by Lake Tahoe, on the California border. I hear it's magnificent. Beautiful streams, lakes, mountains, valleys . . ."

"The company plans to open the mine in late summer," the bishop told them. "My old college schoolmate, Gabriel Stone, will be heading up the operation. He leaves in June, but you'd be getting an early start. You're expected to be in Liverpool on a boat to New York next week."

Tom swallowed. "I . . . don't know what to say."

"How did you do this, Bishop?" asked Annie.

"I learned more than how to blow up things from Thomas." He leaned forward, lowering his voice. "I'm not telling anyone you're testifying, Tom, and we can set it up the day before you have to leave. You need to keep your head down until then. Be careful, and cross your fingers there is no dirty business until I call you to the stand next Monday afternoon, then you could slip out of town early Tuesday morning. The verdict won't be in for at least another three or four days after, and by then you'll be safely gone. They wouldn't dare pursue retaliation until after the judgement is in. It would be suicide for their case. Is a week enough time to make arrangements to leave for Liverpool?"

"We're all ready to go—packed and arrangements made."

"Really?" he responded in surprise. "Well, good then. I'll tell Gabriel you'll both be on board the packet ship leaving from Liverpool in ten days' time. He'll settle his affairs and join you in Virginia City a few months later, sometime in early summer.

Bishop Walker, wise well beyond his years, quietly sat back and studied the two of them. "Thank you, both, for taking up the charge to help these families."

Annie's eyes brimmed. "Thank you, Bishop . . . for everything."

"My pleasure. We'll see you bright and early in the morning, Thomas. We can begin work on preparation for your testimony tomorrow night. You be careful—both of you."

⤝ • ⤜

At the insistence of Bishop Walker, Tom did not attend the opening session of the inquest. "Not until your day in court. It's too risky to be seen in public." So, Annie took the wagon and provided transportation to some of the local miners' family members who wanted to attend. She and the rest of the families were sequestered in the upper galleries, well above the Board of Inquiry, the gathered newspaper reporters, and the rest of the courtroom filled with dignitaries.

Annie looked across the open courtroom to Lord Fitzwilliam's private pew, also stationed well above the courtroom. His Lordship sat, chin in hand, looking down on the proceedings with squinted eyes of disgust, his expression cold and disdainful. She knew that in the days to come the weight of his Lordship's immense power would come bearing down on her husband for what he was about to testify to. The thought sent a chill skittering down her spine.

Reverend Day, Lord Fitzwilliam's personal minister, opened the inquest by declaring, "The entire tragedy was an unavoidable act of God that could not have been helped." He referenced the cost to the mine operators, prominent landlords, and businessmen in London's financial district who had "suffered agonies of spirit" and been forced to accept this "great financial loss."

The raucous crowd of family members in the upper gallery laughed out loud in derision.

"*They've* suffered?" a widow from the gallery called out. "*They're* forced to accept this casualty? These fine lords and ladies have left us widows and orphans to starve in the dead of winter." A grumble of support for her comment swept through the packed gallery.

Coroner Taylor slammed down his gavel. "You are out of order, madam. We will have none of that in my courtroom. Another outburst like that from any of you in the gallery, and I'll have it cleared."

"May I say to the Honorable Reverend Day, your audacity astounds me," Bishop Walker declared. "You say the proprietors of the Oaks Colliery have 'suffered crushing sorrow and great financial loss as a result of an act of God.' You actually claim the insurance proceeds are not near sufficient compensation for the landlord and mine operators' terrible loss. Indeed, I speak in outburst against my fellow cleric, who expresses sympathy for his benefactor and the proprietors of the Oaks Colliery; 'for their crushing sorrow at the terrible loss of life and financial bereavement.' This does not compare to the families who have lost husbands, fathers, and sons. Where are the insurance proceeds for them to survive starvation and homelessness? Those payments have gone to pay for the irresponsible acts of these mine operators, who neglected to install proper ventilation, nor implement the recommended safety measures which would have prevented this horrendous tragedy."

Bishop Walker looked up at the crowded gallery murmuring in anger, before turning back to the coroner. "Your Honor, we will show that these fine aristocratic gentlemen opposed spending money on basic safety measures. Safety measures recommended on multiple occasions. We will also show that their irresponsible neglect cost the lives of over five hundred men and boys under their employ over the last twenty years, even before the Oaks disaster, and still, these operators and landlords were unrepentant for their irresponsible actions.

"If I may, I'd like to call Jenny Lewton to the stand," Bishop Walker requested. Annie watched as Mrs. Lewton nervously took the witness chair before the board, with her littlest one still clinging to her, too terrified to let go.

"Mrs. Lewton, could you please tell us in your own words what happened to your family on December 12, 1866?"

"When the pit went up like a volcano, me and me wee one here was startin' breakfast," began this stoic mother. "All us mams emptied from our houses into the bitter-cold streets and run toward the mine. Wives, mothers with babes in arms, dragging toddlers, heedless of obstructions in our path to the pit. But all we found was cages blown away, smoke, flames, burning wood, rubble, and charred remains of the bodies of our loved ones."

"You lost your husband on that awful morning."

"I did."

"And after a nightmare of desperately searching in the freezing cold, your babe at your breast, what else did you find had happened to your family?"

Tears welled up in Mrs. Lewton's eyes. "I lost seven of me eight children. All buried down in that cursed hole with their papa. All me family gone, exceptin' me wee one here. She's been mute ever since."

"I'm so sorry." Bishop Walker paused for a long moment, giving her time to collect her emotions.

"I died that day in the darkness when me family was taken from me, and me house left in silence."

A chill flowed down Annie's arms to her fingertips. She saw the lonely desperation that had encompassed this poor woman's empty life.

"And what assistance have you received from the company mine operators at the Oaks?"

With disgust, she looked at Mr. Dymond, then up at Lord Fitzwilliam. "Not a farthing offered from those sitting on their perch up there, not even a place to bury me dead family, nor for me and me daughter to lay our weary heads for the night. We been jumpin' from house to house. Only thin' we got from them is eviction notices 'cause me man and boys is dead."

"I'm so sorry, Mrs. Lewton," offered Bishop Walker.

"It's the same fer all of us. Our men are dead, and we been left in the streets to starve! Their Candyman takin' advantage of our young girls livin' in isolation, threatenin' to brutalize the families left behind for complainin'!"

"That's quite enough, Mrs. Lewton," the coroner slammed down his gavel. "You may step down, please. Mind your outbursts, or you run the risk of contempt charges!"

An angry rumble of voices rained down from the gallery and derision rumbled through the courtroom.

Her eyes ablaze, Jenny Lewton stared coldly at the coroner. "Contempt?" She laughed. "Da ya think ya can frighten me with charges of contempt, sir?"

Sympathetic indignation raised up Annie's hackles.

Chapter 32

> ✣ · ✣

BY THE FOLLOWING SUNDAY, EVERYTHING was in place for Annie and Tom to depart for the Liverpool docks. Tom would testify late Monday afternoon followed by a farewell dinner with family on Monday night. Well before sunrise on Tuesday morning, they would leave South Yorkshire forever. The final arguments in the inquest were not expected for several days after. By then Tom and Annie would be on a boat to America.

Annie made one final visit to her little one's gravesite in the potter's field just beyond the church cemetery. Her heart broke again as she stood before the marker she and Thomas had placed over the summer, with the simple phrase, *Henry Wright, may he sleep in peace.*

"My life will never be the same," she whispered. "You'll always be my little Henry."

The pain was too great to linger. After only a few minutes, she turned to leave. Walking somberly from the potter's field, she passed through the cemetery. She saw a tall figure standing in the distance, facing away from her. With his head down he did not see her, but she recognized him at once.

"Thomas?" she whispered in confusion.

Slowly, understanding wound its way through her defenses. Her husband was not here to see their little one. Annie, walking on in silence, stared at him standing before Lydia's headstone.

Not wanting to be seen, she ducked under the chapel portico. Her stomach lurched as she watched him bow his head in tones of solemn conversation. She could not defeat the pain of the twisted blade that was jammed into her broken heart. It seemed to cut into her soul. The disappointment was suffocating, the anguish unbearable.

Lost in her quiet embitterment, Annie became aware of someone speaking to her.

"Madam? Are you well? Can I help you?" She turned to the parish priest. Her hands sweating, her face broke into an embarrassed blush of scarlet.

"I'm so sorry, Father. What did you ask?"

"I couldn't help but notice you observing that poor fellow at the grave. Do you know him? He comes here often, obviously in pain."

"No! No, I don't. I'm sorry, Father, but I must go." Feeling the guilty observer, she found it difficult to respond. Quickly, she strode off without looking back, careful not to be seen by her husband.

How can I speak to a priest about what I'm feeling? she agonized, wiping at her moist eyes with the back of her sleeve as she walked through the cemetery gate. Her heart was shattered, her mind confused. She struggled to take control of her erratic breathing. Her long, strong body, again heavy with child, swaying in a torrent of emotions. Her dark eyes blazed with grief. Her hands shook with humiliation.

"I've been suffering in silence not to burden you with my grief . . . but this is too difficult," Annie anguished under her breath, a weariness shackling her limbs with chains of desperation. "I've been grieving for the loss of our son . . . I desperately needed you."

All the many months she had endured the distance, the silence, coping with the grief alone. She felt demeaned. "Perhaps we are strangers to each other, living in ignorance of our marital trust and commitment?" She paused, feeling as her heart split in half. "Am I a wife who fails to inspire the fidelity of my husband's heart? I have known all along that theirs was a love like no other." But to come face-to-face with it when suffering such loss was too much. "How can I compete with her ghost?"

Inside the safety of their barnyard home, she bent over their bed and sank to her knees. "God, help me bear this," she cried out. "I must put it out of my mind. I must not think about it, or I will burst."

⋟ · ⋞

The inquest began its second week of testimony with a full house. The newspapers had carried the story of the whole salacious affair on the front page every day for over a week. Not only was the courtroom filled to capacity, but the gallery, hallways, and rafters too. There were so many wanting to hear the scintillating testimony that the curious were stacked in the hallways like sardines in a can, up against the chamber doors like snowdrifts, and stacked outside on the porch begging to get in.

"If I may, I'd like to call Mr. Thomas Wright to the stand," Bishop Walker addressed the Board of Inquiry. The courtroom buzzed, sharing shocked looks and comments on this surprise witness. All in the courtroom watched intently as a tall, stately, well-dressed Tom strode up through the crowd to take the stand.

"Mr. Wright, you were the Engineering Manager for the Silkstone & Elsecar Coalowners Company for a number of years, and just recently Engineering Agent for the Oaks, were you not?"

"I was."

"Can you tell us what caused the explosion at the Oaks?"

"Well, the tragedy likely began with a simple accident that might happen on a typical day in mines all across England. Perhaps a glass cover on a miner's lamp broke, and the lighted wick from the oil lamp—or maybe it was a candle—dropped to the floor. Unfortunately, the vein of coal running through the Oaks, the Elsecar-Woodhouse mines, and several other adjacent mines on the same seam of coal is notorious for firedamp. Highly flammable, methane-enriched gases that seep from the freshly cut wall of coal. Like most of the mines along this seam, the Oaks has a woefully inadequate ventilation system. If just one candle or oil-wick flame dropped on the floor at the wrong time in front of the bleeding coal face, it could detonate these seeping gases as it did at the Oaks. When the chamber exploded, it was like the ignition spark in the combustion chamber of a flintlock rifle."

Tom paused to ensure all understood the picture he'd drawn. "The resulting blast at the Oaks traveled from the miner's chamber, through the drift of chutes connecting it to the next chamber, then on to the next. In just moments, the explosions traveled through the web of tunnels, chutes, and drifts, ending in a crescendo of explosions that collapsed the layers of tunnels, one upon another. Delayed chamber explosions continued on for days and even months as the fire burned through the mine and all life was squeezed out of most every soul caught under the crushing weight of fifty tons of earth, rock, timber, and coal."

Bishop Walker stood quiet for a long, solemn moment until the full impact of Tom's words had sunk into the minds and hearts of all those in the courtroom. "Mr. Wright, can you please tell us the difference between death by the firedamp and death by afterdamp?"

"The lucky ones are killed by the crushing blast from the firedamp gases. Death by afterdamp is much more insidious. The breathable air is sucked out by the firedamp blast, and afterdamp death comes slowly by asphyxiation and carbon-dioxide poisoning. It's agonizing. More than a hundred men and boys died this way at the Oaks, struggling to catch their breath, suffocating to death."

Bishop Walker nodded. "Thank you, Mr. Wright. How does the company address this very dangerous problem at the Oaks—and with other mines along this highly dangerous seam of coal?"

Tom looked at the Board of Inquiry, then Lord Fitzwilliam, Thomas Dymond, then long and hard at all the rest of those listening in the stone-cold quiet of the courtroom. "Children."

"Children?"

"Little children, scared—no, *terrified*—are sent down into the mine in pitch blackness. It's so dark they can't see their fingers in front of their faces. It's wet, muddy, awful-smelling coal dust and bodily excretions everywhere so thick you can taste it. Their only constant companions are rats scurrying around in the darkness, some almost as big as the children themselves. These children are assigned to operate the ventilation gates at the risk of men's lives." Tom paused for what seemed like an eternity, then continued in a loud, rising voice. "They send little children terrified at the enormous responsibility. Little boys who sometimes fail in their responsibility. And these little boys died along with the miners whose lives they were charged with saving." Tom paused again, trying to calm his emotions. "It's a terrible thing for a child to live his days in fear. I was one of those children . . . and I will never forgive myself for making a mistake that cost the lives of my friends. I thought I would never get the smell of death out of my nostrils. And over the years, I've paid the price for that mistake many times over."

"I'm so sorry, Mr. Wright." Bishop Walker paused. Another long silence allowed all in the courtroom to digest what they had just heard. "Would you mind sharing the history of similar disasters at the Oaks mine and the collieries along that same vein of coal?"

Tom looked down at his notes. "This seam of coal has had a dozen major disasters over the past thirty years, prior to the December Christmas disaster at the Oaks. On July 4, 1838, twenty-six boys and girls, ages seven to fifteen, were killed in the Silkstone Colliery working that same seam of coal. Despite the horrific conditions and lack of safety standards at the mine, it was deemed an unfortunate accident with no compensation due the families of the lost children. As an eight-year-old boy caught in that tragedy, I was laid up with my injuries for many months." Tom paused with a lump of emotion in his throat at the awful memory.

"Please take your time, Mr. Wright. Continue when you're ready."

"In 1845 six men and boys lost their lives from a firedamp explosion and afterdamp asphyxiation at the Oaks; in 1847, seventy-three men and boys were killed, again firedamp and afterdamp at the Oaks; in 1849 seventy-five at Darley Main on the same seam; in 1851 twelve more were killed at the Oaks, again for the same reason; in 1851 fifty-two were killed at Warenvale. In 1852 twelve more killed at the Elsecar Colliery, firedamp and afterdamp. In 1857, firedamp and afterdamp killed 189 men and boys on the adjacent Lundhill Colliery, and fifty-nine at Edmonds in 1862. These avoidable disasters along this seam of coal occurred mostly because of poor ventilation. With these 504 coal-miner deaths, even before the 1866 catastrophe at the Oaks, deplorable safety deficiencies existed."

All eyes watched Bishop Walker walk across the front of the courtroom, formulating his next question. If I added that up correctly, 504 men, boys, and young girls were killed along this seam because of deplorable safety deficiencies, even before the 1866, Christmas Disaster at the Oaks." Bishop Walker paused, looking at all in the courtroom. "And what changes were made by the mine operators and landlords to ensure the safety of the miners after these disasters, Mr. Wright?"

"After each of these disasters, safety measures were recommended, often including additional ventilation, using more efficient equipment. Specific maintenance recommendations were proposed by the renowned engineer Parkin Jeffcock and supported by the Oaks' pit manager."

"What happened with those needed safety improvement recommendations?"

"Most all were ignored by the mine operators and landlords."

"I object to this line of questioning," argued Reverend Day. "If we wanted a real expert, the so-called renowned Mr. Jeffcock should have been called to the stand."

"I'm sure Mr. Wright would support having his good friend Parkin Jeffcock on the stand instead of himself," Bishop Walker responded. "Unfortunately, Mr. Jeffcock's illustrious career was cut short when he was killed with twenty-five other rescuers in a second blast at the Oaks. He and Mr. Wright, who was also severely injured, were trying to save the miners still trapped in the mine. And, of course, the Oaks pit manager has been forbidden to testify by his employers."

"Please, go on with your questioning of the witness, Bishop Walker," the coroner sheepishly directed.

Bishop Walker turned and nodded to Thomas. "You were acquainted with many of the miners and rescuers, were you not, Mr. Wright?"

"I was. I worked with many of the miners who died. I had long relationships with most of the rescuers—good, honest, capable, and caring men who had come to save lives."

Angry outbursts rippled through the crowded hall. One young woman's voice echoed from the gallery, "And my husband was one of 'em!"

This response spurred the coroner to slam down his gavel. "We will have no further outbursts or I'll clear the gallery."

Reporters scribbled furiously in their notebooks.

"How much money have these mine operators and landlords made over the past few years?" Bishop Walker asked.

Tom scratched his head. "I don't know exactly, but it would be in the millions—the cost of the safety measures a pittance by comparison."

A grumble rumbled through the crowd.

"I have sympathies for these masters of men," Bishop Walker said, striding across the courtroom while he spoke. "But the most painful regret they should surely have is not ensuring the safety of these miners and their families. For the failure of the operators to sink the recommended shafts for ventilation. The evidence I hold in my hand, generated by Mr. Jeffcock before he was so brutally killed, makes it agonizingly clear that if just two of the ventilation shafts had been sunk, this disastrous calamity would never have happened. The thirty years of records spoken of repeatedly tells the story of disaster, followed by warnings, followed by inaction. Sometimes in violation of the law, but always in violation of common decency. These landlords and operators ignored the very measures that would have preserved the lives of those they were responsible for."

Bishop Walker turned back to Thomas. "And why is that, Mr. Wright?"

"For reasons of expense, of course."

"And who are the landlords of all these operations up and down the largest seam of coal in all of England? The men who stand to make the most from the sale of coal and abuse of the miners who dig it out of the ground?"

"There are several. Thomas Dymond of course, the primary operator for the Oaks. Lord Fitzwilliam is by far the largest operator along the seam, with land holdings of over 85,000 acres."

Bishop Walker turned to the courtroom. "Why is it that those sitting in judgement here today, in the most devastating coal mining disaster in the history of England, all seem to be either in the employ of Lord Fitzwilliam or have some other obligation to him?"

The coroner slammed down his gavel. "Bishop Walker, how dare you. That fact is irrelevant to this inquest."

Bishop Walker looked up at a disdainful Lord Fitzwilliam, sitting in judgement well above the fray in his private pew, coldly unmoved by the testimony of the miners' plight and his part in it. "And why do you think His Lordship has refused to allow the Oaks' pit manager, Mr. Tutor Turton, the most knowledgeable man still alive on the Oaks, to testify before the Board on his efforts to institute further safety measures over the past many years?" Without waiting for an answer, Bishop Walker continued. "For a few pieces of silver, these masters of men traded in the lives of these desperate families." Bishop Walker was no longer speaking to the Board of Inquiry, but to the press, the politicians, and the public across the whole of England. "Three hundred and eighty-three men and boys dead. Thousands more maimed, and tens of thousands of lives torn apart because these 'masters of men' did not heed the repeated recommendations of their own pit manager and his technical advisors after hundreds had been slain." He turned and directed another question toward Tom. "How could they?"

"They're aristocratic industrialists," Tom responded matter-of-factly. "They are not subject to the same laws you and I are."

"As I understand it, you had a relative who died in the Oaks Mine disaster."

"My cousin, Will Wright."

"Were you close to Will and his family?'

"I was best man at Will and Emily's wedding. My wife and I were living with them when the disaster occurred. My wife assisted the midwife in delivering Emily's first baby just three weeks after the Oaks disaster."

Bishop Walker nodded. "Will had just returned home after being incarcerated for over three months, working on a chain gang at the direction of Lord Fitzwilliam's agent—why?"

"Will had the temerity to complain about excessive firedamp at the Oaks. He and his fellow miners, 'conspirators' they were called, were prosecuted and sentenced to hard labor for disturbing the peace. Of course, Will and the others were absolutely right, but he embarrassed the wrong people, and his family suffered for it. Will had just finished serving his sentence when he went back down into the Oaks Colliery. He was working double shifts to pay off debts incurred because he was unable to earn a living for his family while incarcerated. And he wanted to make a little extra Christmas money to buy a special present for his wife and their new baby.

"What happened to Will?"

"Will suffocated to death in the afterdamp. I was there when Emily and her brother found him lying dead in the yard outside the Oaks canteen. I can only imagine what those awful, final minutes of his life were like."

Bishop Walker paused to let Tom's words sink in. "I'm sorry, Mr. Wright. Thank you for your testimony."

The courtroom was silent, except for the reporters' frantic scrawling. *The London Times, Daily News,* and newspapers from all over the United Kingdom were represented; the list had grown longer every day as the appalling story of the "Oaks Christmas Disaster," "The Great Golgotha," caught hold of all the United Kingdom's attention.

Tom felt a chill run down his spine as he watched this self-obsessed, heartless Lord Fitzwilliam grip the railing in front of him. With a contemptuous flick of the wrist, Lord Fitzwilliam called over an attendant, whispered a message in his ear, then sent him on errand to the coroner conducting the inquest.

There was a shuffling of papers, shared whispers between the board members, then the coroner drove his gavel down hard on the dais. "It is the decision of the Board of Inquiry," the coroner declared, "that no further testimony is warranted. Reverend Day and Bishop Walker, please make your closing summations."

This surprise decision caused a rumble through the courtroom and a firestorm of outrage in the gallery seated above. The newspaper reporters, who now flowed into the hallways, spoke in confusion among themselves, aghast at the unorthodox decision to cut short the prosecution's testimony.

"In conclusion," offered Reverend Day, "this country can't risk slowing down coal production just because we were unfortunate enough to be confronted with 'an act of God.' Coal is the lifeblood of forward progress for this country . . ."

"That may not be the only unfortunate act of God that His Lordship faces," Tom, now sitting in the gallery, whispered under his breath. But even he was surprised and appalled at this travesty of justice.

"What price can be placed on the lives of these hundreds of human souls lost because of the irresponsible acts by these drivers of industrial slavery?" Bishop Walker began his closing statement. "And the thousands more left broken, widowed, or orphaned. What price should be put on the families left homeless without enough food to eat or sufficient finances to even bury their dead?" He paused to let the question sink in. "These industrial slavers have brushed aside their responsibility for the dead and their families left behind to starve. As

Queen Victoria put it when speaking of these aristocrats, 'May God have mercy on their souls.'"

Bishop Walker, who had held the courtroom spellbound, concluded with a quote from England's prime minister-elect, Benjamin Disraeli, on the Oaks Christmas Disaster, who paraphrased the words of Montesquieu. "'There is no greater tyranny than that which is perpetrated under the shield of law and in the name of justice. The only difference between these aristocratic industrialists and cannibals is that cannibals don't eat their own.'"

The raucous miner families and newspaper men pounded their feet on the wooden floor, furiously calling for justice against the irresponsible acts of the mine operators and the landlords who benefited. The reverberation through the hall was deafening.

The coroner announced, "We'll take a short recess, then be back with our verdict." He slammed down his gavel, and the members of the Board of Inquiry retired to their deliberation chambers.

Tom knew the scale of imbalance was tilted heavily against the miners, no matter the facts of the inquest. These wealthy operators and landlords held all the cards. The bishop expected a short deliberation in this sham inquest, which would almost certainly weigh on the side of these aristocratic industrialists.

The families in the gallery awaited their fates in silence, hoping for at least some compassion for their plight.

"How does anyone live through this?" Tom muttered.

The clerk entered the room, calling all to order and announcing the Board of Inquiry had reached a verdict. The room, filled to the rafters, fell silent. Many were bunched in the hallways, on the steps, hanging on the railings—anywhere there was space to stand, overflowing into a crowd outside awaiting the outcome.

"All rise," the clerk called out.

Like a funeral procession, solemn and cheerless, the board members filed into the chambers and took their seats. One of the members glanced at Tom, others at the family members in the gallery. But the eyes of most never left the floor. No one in the courtroom made a sound as the coroner prepared to deliver the verdict. Tom felt nausea rise in his stomach as he tried to read the coroner's face. He sensed a collective breathlessness of all in attendance, their hearts pounding as one.

"We, the Board of Inquiry," read the coroner, "find it unnecessary to make any special recommendations as to the workings of the mines. No damages are to be awarded to those injured by this act of God, nor compensation to the families of those unfortunate miners killed in the course of their work."

One of the wives with four little children at home swayed where she stood next to Tom, her knees buckling. Tom reached out and caught her. He held her up, pulling her close. Another fainted, falling into the gallery railing.

"How could you?" some cried out in anger.

"This ain't justice!" others echoed in disgust.

The outbursts of indignation in the chambers echoed in the rafters.

The coroner hammered his gavel. "We've heard just about enough. This hearing is over."

But no amount of pounding of the gavel seemed to make any difference. No one left. The sound of stamping feet and chants of injustice rumbled through the wooden floors and reverberated off the walls.

Tom, standing motionless, blinked then let out a long-held breath. It was the verdict he had anticipated, but still—to be insulted with the words thrown in his face was almost more than he could bear.

The London newspapers would call it "A travesty of justice by powerful men."

Tom could see the indignation written across an unrepentant Lord Fitzwilliam's face.

✢ • ✢

Lord Fitzwilliam looked around at the plodding peasants who had come to gawk at their betters, his stomach churning as if he had been drinking sour wine. He hated the old men, the broad-hipped women, the apple-cheeked children who sat in the rafters looking down on him. He was the Lord of the manner, after all, and he detested these grasping, dirty, uncouth commoners for confronting his position of authority. Where was their gratitude? He had done his duty to support them, to provide shelter and work for those not too lazy to put in an honest day's labor.

"The people?" he muttered with disdain. "They will pay. All of them."

But it was the betrayal of Thomas Wright and his family he resented most. They had been a thorn in his side for far too long. He burned with bitterness and hatred, and found the anger that brewed cold inside him nearly impossible to abide. Tom Wright had been responsible for his public undoing, for the blight on his reputation. He was responsible for casting scandal on his betters. Who was the Wright family to make him a spectacle in front of the press, to make the Fitzwilliam name an object of ridicule all over London and across the United Kingdom?

The son is like his mother, he thought. *She was a haughty little tart thirty years ago, and that family has been a plague on my house ever since. We should never let them learn to read. Look at him. The audacity—to meet my gaze so directly.* Lord Fitzwilliam glared back at Tom with cold contempt, the author of his humiliation, the object of his cold, vengeful ire.

Gripping the railing in front of him with white-knuckled rage, he moved his icy stare purposefully away—first to Bishop Walker, then to the courtroom busied with scurrilous press. When he turned back, Tom's attention had been drawn elsewhere, and a bitter Lord Fitzwilliam muttered his final benediction. "Watch ye therefore: for ye know not where or when your master cometh *for you*."

Then, a single nod to a waiting Candyman Boo Black standing in the chamber doorway directly across from His Lordship's private pew. Boo lifted his head high. A chilling smile of acknowledgment crept across the evil man's purplish, perspiring face, exposing his two silver front teeth. The pulsating vein on his sweaty forehead rose as his bloodshot eye filled with a sinister gleam of satisfaction darted toward his unwitting victim.

Chapter 33

≯ · ≺

A
T THE FRANTIC INSISTENCE OF a hysterical Fanny Templeton,
Tom stopped by to see her on his way home from the inquest.
Fanny, the wife of one of the dead rescuers, was holed up inside her
neighbor's shanty after being badly beaten by Boo Black for answering questions at the inquest, then thrown out into the streets.

Tom tried to reassure her. "It took a lot of nerve for you to speak
out, Fanny, after losing Michael in that awful tragedy."

"Me Michael dropped everythin' to take up with you in rescuin'
those poor boys trapped at the Oaks. 'Twas the right thing to do. But
now me three little ones're starvin'." She wiped the blood from her
forehead. "He told us to get out . . . and never come back," she stammered. "Or next time, he won't be so gentle with his fists."

"I'm so sorry, Fanny," Tom whispered. "Mr. Jensen and his wife
are watching your children. They're feeding them a little dinner. The
Jensens have agreed to take your family in for a few days. That seems
the best we can do until we figure something else out."

Fanny sniffled. "Thank ya, Mr. Wright, and your angel of a wife
too."

≯ 340 ≺

"Here's a few shillings to tide you over." He reached into his pocket, pulled out all he had, and handed it to her. "It'll be all right."

There was a hard knock at the door. Mr. Jensen opened it. Tutor Turton stood in the open doorway, bathed in the light of a lantern.

"I apologize for barging in on your home this late, Mr. Jensen." Tutor wiped the sweat from his brow, anxiously looking up and down the alleyway. "I was told I might find Mr. Wright here."

"That's quite all right, Tutor."

Tom slid in beside Mr. Jensen.

"Can we speak privately for a minute, Tom? Outside?"

"Sure."

Both stepped out on the porch, and Tom closed the door behind them.

"It's a good thing that your doin' here, Tom."

"Thanks, Tutor. But what choice do I have when they insisted I come? Poor folks."

Tutor hesitated for a moment, trying to catch his breath. "Sorry, I run all the way here and my heart's poundin' fast enough to keep time with the wings of a mallard duck in flight."

Tom waited patiently.

"There are some real upset folks 'bout what ya done at the inquest this afternoon."

Tutor glanced around nervously. Tom nodded.

"It ain't good," Tutor began again. "I'm here to tell ya—you're in a heap of trouble with the company and Lord Fitz. He don't like to be embarrassed, labeled a fool, or worst yet, the 'disreputable evil landlord,' as the newspapers all over London been sayin'. Word is they're after ya. And o' course, ole Boo has gone plumb crazy." Tutor wiped the sweat from his forehead. "He's a comin' for ya, Tom. Ain't makin' no exception this time. I'm afeared for yer family too. Ain't no time to waste."

Tom's heart leaped into his throat. "Oh my God!" How could he have left his family alone? It was a question he would ask himself a thousand times.

"Boo and them's on the rampage with ole Fitz's blessin' to take revenge and then some. They're carryin' guns."

"Any idea exactly what Boo's planning?"

"No tellin'. But I don't 'spect any of ye be wantin' to find out. Best be movin' on."

Tom stamped his foot. "I gotta go!"

He'd already turned to run when Tutor grabbed his arm. "Here, take this. It's them papers ya been askin' fer." Tutor handed him a large valise of documents. "I'll probably regret it in the mornin', but after what I heard this evenin' . . . just be careful."

Tom shook his hand firmly. "You're a good man, Tutor. It's a hard thing you're doing here."

"It's the right thing to do, ain't it? What choice do a man have if he's gonna be facin' Saint Peter someday? Probably sooner than I'd hoped."

Tom stuffed the valise of papers into his knapsack and pulled it onto his back. Then, with dread coursing through his veins, he began the long eight-mile run toward home.

⊁ · ⊰

All the women in the Wright family had gathered in the kitchen, each taking up their assigned tasks from Martha for the farewell dinner. Chatting and laughing together, they went about their work on their last night together as a family. Joey and Andrew were out chopping wood. The wagon had been loaded and the horses fed, ready for their trip to Liverpool in the morning. All the children, unbothered by the cold, were playing in the yard, twilight settling in as dusk began to fade on the cloudy winter evening.

"I can't believe it's finally going to happen," Edie sighed, pausing from cutting up two plump rabbits Thomas had brought home the night before. "I can't imagine it. I'd be shakin' in my boots if I were off to America." She wiped her wet hands on her apron, then picked up the bacon. "I'm gonna miss you somethin' terrible, Annie."

Living so close together in the family compound, Edie had become as good a friend to Annie as a sister could be. Usually quiet, still she was determined, organized, and disciplined in her demeanor. Her tired brown eyes, permanently rimmed in scarlet, came from keeping her five children under her protective wings. For all her plainness of feature, there was a sedate dignity about her, a trustworthiness in her bearing as solid as gold. In short, a hardworking mother.

Annie pulled down the spices off the shelf. "I wish you were coming, Edie."

"Oh, Annie," Elizabeth, Edith's pretty and unusually precocious twelve-year-old, interjected. "Aren't you excited?" Working in tandem with Annie on the mashed potatoes and greens, Elizabeth was the only one in her family who ventured far from her mother's skirts. She held everything around her in fascination. "I can only imagine the great adventure of it all—crossing the ocean and rivers, meeting Indians, wagon trains crossing the plains!"

"I've always wanted to travel," Martha chimed in, "but I was thinking maybe London. I can't imagine going all the way across the Atlantic Ocean to another country."

"Not that London ain't another country," Annie interjected as she looked out the window at the gathering storm clouds. "It looks like the drought might be over. I sure hope Thomas doesn't get caught in the rain. Where is he?"

"Thomas will be careful," Martha spoke boldly, trying to hide her own concern. She didn't put it past the company to set him up for trouble, even before the decision came down from the inquest. "I'm sure he's fine!"

"Ol' Fitz will be on a rampage," Annie said, lowering her voice in concern. She absent-mindedly finished cutting up the first of the two rabbits for the frying pan. "Thomas doesn't say much, but I know he's afraid the company has it in for him. Honestly, his plan to leave in the middle of the night seemed a bit cautious, but I can't wait until we're out of here for good in a few hours. I just hope he's careful and gets home soon."

"It's a scary time," Edie agreed fretfully. "Speaking truth doesn't come without consequences."

Suddenly, the two solid-oak front doors burst open with a crash, the sound of broken glass splintering and metal clattering in the front parlor. The women in the kitchen spilled out into the parlor, questions coming in a flurry.

There were muffled shouts from outside, then Candyman Boo Black filled the doorway. "You men," he yelled out his commands, "drag those two inside. Get those little nits in here too."

Four of Boo's thugs emerged, manhandling Joey and Andrew, pushing them roughly into the parlor. Andrew tripped and sprawled face-first across the floor, cutting his hand on a shard of glass, bleeding all over the floor. Edie rushed to him, and the terrified children huddled around their papa, crying.

"Shut them up," Boo demanded.

Elizabeth scooped up the youngest two, and Annie wrapped her arms around the others.

A seething Boo Black swung his flintlock rifle in a wide arc as he raged around the room, fire in his bulging, bloodshot eye, spittle on his unshaven chin.

"Where's Wright?" he spit out in a drunken slur.

"How dare you burst into our home and accost my family, Mr. Black!" Martha defiantly stepped forward to face their tormentor. "You're scaring the children."

"Where is he?" Boo lunged toward Martha, grabbing her by her throat, and pushing her up against the wall. "Yer family's caused me enough trouble."

Joey rushed to defend his mam, but the butt of a rifle caught him under the chin, and he came crashing to the floor in a heap.

Martha wrenched herself free of Boo's grasp. "You filthy animal," she rasped. "Who gave you the authority to come into my home and harass my family? To brutalize my sons and terrorize these little ones?"

Boo turned his flintlock on Martha, putting the muzzle to her forehead as she stood tall and erect. Her back against the wall, she was forced to stare into the singular maniacal window of Boo's deranged mind.

"By my authority," he growled. "You Wrights think you're somethin' special. Well, it's time we get rid of ya—once and fer all."

Martha spit in his face.

Andrew scrambled to get in between them, but Boo slammed the cold hard steel of the barrel across his face.

"No, Andrew!" Edie screamed as blood gushed down his cheek. "Please, don't hurt him."

The thugs behind Boo Black slammed a club into Andrew, knocking him back to the floor. The children wailed in terror, blood splattering everywhere.

Annie, trying to control her panic, shouted from the floor where she held the children, "Thomas is not here!"

Boo, his face a cardinal red, turned toward her. His eyes blazing, he put up a hand to stop his men. He put his free hand under Annie's chin, lifting up her face, forcing her to look into his bloodshot eye. He drew out his words slow and calm, in a way that chilled the bones of all in the room. "This here family has given aide to Thomas Wright," he sputtered. "You've conspired against your betters for the last time. The hand that feeds ya has had enough."

There was only silence in the room except for the whimpering of little children huddled around their mam, Elizabeth, and Annie. Joey and Andrew, bloodied, arms wrenched behind their backs. Only Martha remained standing straight and strong, resolute as she stared back at Candyman Boo Black.

"You can go to hell, Mr. Black," she answered coldly.

With a contorted taciturn scowl and a countenance as cold and maniacal as Satan himself, Boo drawled out his final response. "Well. No matter." He stretched to full height, sneering at all in front of him. Without further threat, he turned and walked toward the door.

"Lock 'em up," Boo muttered to his gang of thugs as he walked out into the yard.

"You're scum," shouted Joey. "You'll pay for this!"

"You'll have to take that up with your Maker," Boo coldly retorted without turning around.

The doors slammed shut behind him. Those inside heard the chain slide through the iron handles and the lock click. Boo ordered a heavy timber to be nailed across the heavy oak front doors. The oak shutters on the windows were slammed shut and barred.

The family listened as the house was battened down.

"Open this door!" Martha demanded.

"Burn it down!" Boo shouted from outside.

"Fire the barn, Mr. Black?"

"The barn? Oh no—the house," Boo Black calmly insisted, his brows raised, evil eye vacant.

Andrew pounded on the door as the children cried out, terror-stricken and in tears.

"But they're all inside!" a man cried out in shock. "Tom Wright ain't even here."

"He'll be here soon enough," Boo growled. "Then we'll kill 'im too. But not afore he learns his pregnant wife, the old woman, and the rest of 'em has been burned alive."

"I cain't do that, Mr. Black."

"Hand it to me," Boo snarled.

There was the shuffle of footsteps, followed by a soft thud on the front porch up against the oak doors. Annie was the first to notice the whisper of smoke drifting up from under the two front doors. She looked at Martha in disbelief—had Boo Black actually lit the house on fire?

She could hear the crackle of the fire beginning to burn outside in the dry weeds. Within moments she could taste the smoky ash that began to swirl about the darkened room. The children were screaming. Annie coughed. Her chest hurt as she drew in hot, short, shuddering breaths. Trying to hold down her emotions in the rising heat, she took shallow breaths through tight lips.

Edie's youngest, two-year-old Tommy, began to cough violently. Annie wrapped her arms around him and pulled him into her. Sweat was beading up on his weary brow, his little heart pounding in her arms. She sat down with Tommy, beside Edie and Martha, who were holding the rest of the children.

"Help me, Auntie Annie," Tommy cried out. "I can't breathe."

Both Andrew and Joey, using their combined weight, slammed their bodies into the great oak doors again and again, but the heavily barred doors would not yield. Chairs and tables were thrown, crashing through the glass windows, but the shutters held tight. Joey ran to the cellar, but there was no way out, no escape through the door barred from the outside.

The heat bore in on them. From every crack in the rickety old farmhouse, smoke seeped in, filling the room. Andrew guided the rest of family down the stairs, leading them into the cellar. Sweat blurred Annie's vision, ran down her face, her neck, and between her breasts in itchy streams. As she reached for the railing, she stumbled in the dark, smokey room and fell hard to the floor on to her hands and knees. Her badly scraped arms shuddered when she tried to pull herself up.

Nine-year-old Johnny, who had strangely gone silent, reached for her. "I'll help you, Auntie Annie. Here, hold my hand." He put down his open pocketknife he'd been holding in the ready to fend off the bad guys if they came back, then helped his Auntie up off the floor. They carefully made their way down the stairs to join the others sitting on the floor in the corner of the cellar, where the smoke had not yet penetrated.

Martha sat down beside Annie and reached out to steady her shaking hand, squeezing it. Feeling Martha's clammy skin on her own trembling fingers, she held on tight and placed both hers and Martha's hand on her stomach. "My baby! What's to become of my baby?" Annie whispered in a scratchy voice. "I'm afraid!"

Martha's calm voice seemed to sooth her. "I'll be with you, honey."

Glass began to crack and shatter upstairs. Timbers snapped. Heat sucked out the humidity from the air. Annie tried to brush the descending tendrils of smoke from her eyes as they drifted down from above. Taking in short breaths, she tried to ignore the pounding of her heart in her throat.

Edith, who wore her alarm like a shroud, had snuggled in close on the other side of her mam, her arms wrapped around her brood of whimpering children. Dressed in her dowdy frock, her skin was pale and drawn in the flickering candlelight. Tenderly, Martha smoothed the damp hair on her daughter's sweaty, ashen brow.

Andrew knelt in front of his family. He put his arms around his trembling wife and whispered words meant for Edie's ears only. Then he kissed her, and kissed each of his children one last time.

Without a word, Martha slowly stood up from the floor. The light from the single candle in the darkness reflected the fear and hopelessness in each face as all in the family looked up toward their matriarch. As Martha began to pray in a surprisingly clear and serene voice, Annie felt a calmness wash over her, her breathing slowed, her shaking subsided, and her hysteria over her certain fate receded.

"The Lord is our shepherd; we shall not want."

The lifting up of Martha's heartfelt voice to God seemed to give each terror-stricken member of her family the feeling of hope for salvation.

"He maketh us to lie down in green pastures: He leadeth us beside the still waters. He restoreth our souls. He leadeth us in paths of righteousness for His name's sake. Yea though we walk through the valley of the shadow of death, we fear no evil, for Thou art with us . . ."

⊱ · ⊰

Outside, one by one, the dreaded gang of thugs drifted off in shame, disappearing into the dark shadows of the night, not wanting to acknowledge the eternal consequences of their devilish part in the fire. All left the scene of their crime except Boo Black, who stood transfixed by the flames.

The sound of the family's screaming and shouts for help no longer carried on the wings of the hot breeze in the evening darkness.

All had gone silent, leaving only the hypnotic translucent flames, the crackle of burning wood, the awful smell of smoke, and the taste of ash on the hot air. The fire flowed from one side of the aged, termite-ridden farmhouse to the other, the rising heat chasing ash high into the night sky.

Chapter 34

・<

TOM STAYED OFF THE MAIN road as he ran toward the family farmhouse perched high up on the bluff, at the base of the Peak Mountains. As darkness settled in, the uneven pathway carved into the steep cliff edge became difficult to see. Although he knew these woods well, they always looked different at night, even under the full moon. It seemed the afternoon shift of trees, plants, flowers, and rocks had gone home for the evening and the scarier nighttime crew had taken over. Everything had an unfamiliar shadow, cloaked in dangerous hues, shadowed shapes making the rootbound, rocky terrain difficult to navigate.

The farther he ran along the less conspicuous backwoods shortcut, the faster he went. He was desperate to get to his family. His heart quickened as he picked up the pace. He could see puffs of white clouds, his breath hitting the cold night air. His legs, back, and ankles ached. Sweat ran down his face, and his clothes clung to his body.

The absolute stillness of the forest gave the feeling there was something amiss. There were no signs of life in the shadowed darkness, but still he knew there was something out there. How foolish he'd been to leave his family alone—Annie, Mam, Edie, her little ones, all at risk? What had he been thinking? Maybe evicting Fanny had not just been for the villagers' eyes only, to scare them into compliance, but instead a ruse to draw him away from protecting his family. To draw him into a trap.

With his heart lodged in his throat, Tom swept around the last bend heading toward the open meadow and the farmhouse beyond. With the trees thinning out, he was fully exposed to the dangers of the night. The jagged canyon walls cascading down to the valley floor were flooded in moonlight.

He shuddered as he ran. It would be the perfect place for an ambush. It was just like Boo Black to set a trap for him here.

Then he saw it. The soft glow above the tree line. A whisper of smoke drifting high into the darkness. Pulsing fear screamed through his veins. Pushing aside all prudence and caution, Tom broke into a dead run, thundering up the rocky pathway along the canyon edge, pressing boldly on toward the moonlit meadow beyond. Adrenaline surged through him as he dodged obstacles. His aching legs were forgotten, his lungs that had seemed ready to burst in pain ignored, and his desperation to draw in more air set aside.

"Annie!" he screamed out in a ragged voice, knowing anyone who lay in wait would hear, but hoping maybe the distraction would give his family a chance to run. "I'm coming!"

A rifle shot rang out, breaking the silence of the woods. He heard the *crack* of a tree limb as it fell to the ground. He ducked, then dropped to his knees. His chest was heaving, his heart pounding.

His eyes darted from side to side. What direction had it come from?

Crack! Another shot rang out. The bullet whizzed past his head and, with a thud, stuck in the tree next to him. All of a sudden, what felt like a searing, red-hot poker skewered the top of his right ear. He pulled into himself for safety. His heart pounded in his throat. With his hand he could feel the pulsating thickness of blood. The top of his ear had been blown right off by the bullet. A gush of blood poured down his face and neck and into the corner of his mouth, leaving a metallic taste. Another bullet shrieked through the forest, snapping limbs, ricocheting, and echoing off the canyon walls.

Tom bent down below the brush line, his eyes darting from side to side in search of the shooter in the darkened shadows of the forest.

Pain surged through the right half of his face. Blood soaked his shirt and coat.

There was the sound of breaking branches as someone plunged toward him through the brush. Tom spun around toward the sound in time to see the contorted face and blazing eye of Boo Black diving down on him from the slope above. He was wielding a knife that glistened in the moonlight.

Instinctively, Tom turned away and braced for impact. Boo slammed into him so hard, his momentum sent both men sprawling across the ground. Tom felt the heavy thud of the knife blade lodging deep into his knapsack, his face pushed into the dirt. The momentum of the blow ripped the knapsack off his shoulders, the knife still stuck in it.

Both men tumbled on the ground, grappling with each other. Tom was caught in a crushing headlock. The strong, sour smell of ale mixed with the reek of sweat assaulted his nostrils. He found it almost impossible to breathe, his head burning, blood pounding in his ear. He felt a crack, then a sharp pain in his neck.

"I'm gonna twist your head right off, Wright."

Tom's muscles seemed strained to the snapping point.

Fear turned to anger. The adrenaline rush drove his muscles to Herculean strength, and Tom pulled himself from Boo's grasp, pushing Boo to the ground. Tom came at him like a cannonball rifling through a barrel, teeth clenched, and with an adrenalin-riven surge of strength Tom plunged his fist into Boo's face, again and again. He felt the dull crunch of bone, the warm sticky spray of Boo's blood splattering across his own face.

Still, Boo seemed undeterred, pulling free in a wild rage. He grabbed a nearby fallen tree limb, and with immense drunken strength, swung it into Tom's stomach. Knocked breathless, his head swirling, Tom grabbed his stomach in pain. He stumbled back, tripped, and fell backward. Arms outstretched, he crashed through a tangle of brush along the cliff edge. His eyes wide in panic, he careened over the ledge, arms flailing, hands grabbing at the empty air. He bounced off a lonely scrub oak jutting out over the canyon wall, miraculously wrapping both arms around its trunk and hanging on for dear life.

While he seemed to have been saved from falling to his death, now he dangled over the canyon below, four feet from the cliff edge. Panicked and out of breath, blood running down his face, he was too exhausted to struggle up. He hung there completely vulnerable, looking up at the crazed, raging bull of a man. Drunkenly weaving back and forth, Boo brandished his heavy club out in front of him. He wiped at the blood gushing from his crushed face into his gunmetal eye, under thick caterpillar brows. A wild-eyed contemptuous grin spread from ear to ear.

"You stinkin' bastard. You're a dead man, Wright . . . but when I says," Boo sputtered out, spittle mixed with blood trickling from the corner of his mouth. "When I decide ya done suffered enough, I'm gonna bash yer head so hard, nobody'll recognize yer body at the bottom of that there canyon."

"Do it then."

"You shoulda seen 'em. The look on your bitch of a wife's face. All terrified, hands tryin' to cover that little nit of yourin' inside her."

Boo released a sinister laugh. "And your ole lady, whinin' when I put the torch to the house and lit 'er up." He swayed drunkenly, swinging his unwieldy club back and forth. "All of 'em screamin'. Like sows in a pen. We fried 'em like bacon."

Tom arched his neck and, with all his remaining strength, spat blood and saliva at the drunken man.

Furious, Boo raised the club high overhead to deliver one massive, crushing blow to Tom's skull. Hurtling himself forward, the raging bull swung the heavy club down with all his might. Defensively, Tom pulled up his knees, twisting his body away from the pending blow. With one mighty swing the club came crashing down, but it missed its target, and the momentum of the drunken monster sent him sailing past Tom, launching himself right over the cliff edge. Boo Black plummeted down the canyon face, screaming obscenities all the way.

His heart pounding, Tom couldn't believe he was still alive. *That drunken lunatic missed me.*

Boo Black's body lay crumpled in a heap fifty feet below. Adrenaline still pumping through his veins, Tom struggled to pull himself up enough to straddle the scrub oak. Slowly, carefully, feeling his body's pain, he slid himself to safety.

Spread out on his back, legs splayed open, he felt the exhaustion break over him, the breathless pain. He turned and vomited. Reaching deep inside himself for strength, he sat up and stood shakily. In the quiet of the night, he heard the faint moans coming up from the tangled body far below at the bottom of the abyss.

Tom retrieved Boo's horse, untying it from the tree trunk. He picked up the empty flintlock and slid it into its scabbard. He took a long swig from the canteen hanging around the saddle horn, coughing up cheap whiskey, then climbed aboard and spurred the horse into action. The horse whinnied, the bridle jangled, and he galloped off into the dampening night, thundering down the trail toward his family's home.

When Tom reached the farmhouse clearing, he pulled up short, his wild eyes searching the flat hilltop where the farmhouse should have been. His breath caught in his throat. Impatiently, he tried to wipe clear his vision by pulling his dirt-caked palms down his disbelieving face.

"Maybe . . ."

But all was deathly quiet.

He slid off the horse and stood stone-faced in front of what was once his family's home.

There was no sound of children. No hugs from his caring family. No one talking about babies. No sweet smell of a farewell dinner. No caresses from a loving wife. There was no one at all.

All was eerily silent, except for the sizzling mist of a light rain falling on embers, bringing a fog up from the charcoaled timbers and smokey debris. Tom stepped forward into the remains. The house seemed smaller in death than it had as the family home. He inhaled the musty smell of charred wood and smoke, tasted the ash-ladden mist lifting off the remains of all he knew. He tried to sift through it, but the embers made it impossible to handle the remains of his family's home. It was difficult to even step through the debris. Still, he pressed on, mind numbed, looking for clues, anything to tell more of the story.

Boo's disparaging comments rang in his ears. "There couldn't have been anyone still inside," Tom whispered to convince himself. "It isn't possible."

He stopped to steady his unsure legs, to calm his wild imaginations. He looked down at the two great oak doors, pride of the Wright family. They had burned hot in the fire. Embers still traced the intricate lines carved by Papa into the wood; the dull tarnish of the two iron door handles were all that remained. Absently, he reached down to touch them, yanking back scalded fingers. Then he noticed that the handles were chained together. Tom's heart jumped into his throat. "No! It can't be . . ." A suffocating panic seized his heart; he was sure it would pound right through his chest.

"Annie!" he called out. "Mam! Joey! Where are you?"

He paused for a moment.

"Edie . . . ? Andrew? Oh, please, God . . . do not forsake me now!"

He clenched his shaking hands into fists. The acrid smell, the steam, the smoke that rose off the ash left an awful taste in his mouth. It sickened him. He swallowed hard. Nothing seemed recognizable in the shadows of the moon-filled night with ash and soot covering everything.

His knees began to buckle. He stopped breathing. Woozily, he reached for the still-standing stone fireplace to steady himself. His darting eyes searched frantically through the rubble, desperate to find something—anything—to indicate his family was still alive.

He saw the ancient metal tub he had once used as his study table, Mam's favorite cooking pot. Sitting on the concrete steps leading to the cellar was his old pocketknife. He'd given it to his nephew Johnny. It was his prized possession. The cellar was filled to the brim with charred timber that had caved in from the roof and debris that was still burning hot.

Then, glistening in a shaft of moonlight where the wooden bench had once stood by the cellar stairway, Annie's bracelet. A present from Emma, she had never once taken that bracelet off. His throat tightened as he reached down into the ashes to pick it up.

"My God," he cried out, his voice raspy. Lightheaded, suffocating dread seized his mind, impossible thoughts raced through him.

"Oh, dear God, what do I do now? Annie . . . our baby! No! . . . No!" He leaned over and retched again, heaving his insides out.

His thoughts in total chaos, the pain was unbearable. "They take everything. Everyone. Even our wives and children."

Tom pressed the bracelet to his lips, then closed his fist around it. For a long moment, he stood silent, tears running down his face, the lump from deep within his stomach now caught in his throat. Dazed, confused, finding it difficult to breathe, he didn't know quite what to do next. Then, unable to look on the wreckage any longer, he stood up straight.

"I have to leave here, or I'll burst."

Hate raged through him. The injustice, the savagery, the barbarism of this act. He walked to the barn in long, determined strides. Though empty of any sign of life, it seemed to be untouched by the disaster. He grabbed a bottle of rubbing alcohol meant for the horses and doused his arm and ear, jerking at the sting, then wrapped his head with a clean cloth and turned to retrieve his bow, his quiver of arrows. He secured them to the back of the saddle, then mounted Boo's horse. And without looking back, he rode off into the night.

His home and family were gone. His plans for the future were gone. But he wasn't afraid. Blind fury had seized his heart and consumed his soul, leaving but one cold, dark thought: *revenge.*

He hated Boo Black for what he had done, for his sick, blackened soul of sadistic ruthlessness, but it was Lord Fitzwilliam and those like him whom he blamed most. The anger within him began to harden into cold, inflexible determination.

"I promise, if it is the last thing I do in this life, they will pay. They will pay for their sins, and I will join them in hell."

Chapter 35

❧ • ❧

As Tom rode into the night, guilt began to seep into his soul. Why had he not been there for his family? How could he have been so foolish? Had he just been naive to the danger after the inquest was cut short? Maybe he was a coward. The revelation didn't diminish his hatred, his rage for those responsible, but how could he ever claim clemency for his crime of not protecting his family?

The rain started to fall in earnest, its icy fingers running down his back. Traveling high on the hills through the pine, cypress, and spreading oak forest, he kept an eye on the valley below to ensure he saw his enemies before they saw him. But no one came. He plodded on through the night until he was too exhausted to go farther.

As the rain stopped he pulled up to spend the night in the forest. Too distraught to perform even the simplest of tasks, he took a blanket from the saddlebag and bedded down for the remainder of the night beside the road.

Beyond exhaustion, he shut his aching eyes and let his body sink into the forest floor, still listening to the sounds of an awakening forest after the storm. Slowly, he slid into slumber. His mind drifted not into the nightmares he'd expected, but into untethered thoughts of family. Fond memories of his first words reading with Mam; playing in front of the fire with Papa, Georgie, Joey, and Edie; Edie's children filling the house with laughter; Mam's unwavering dedication to right his world; Annie's tender loving-kindness, smiling at each other while they watched the sunset, her hand in his. Without even recognizing it, she had stolen his heart. He twisted and turned in his sleep. Why had he not seen it when she was alive. Why had he been so careless, taken life's blessings for granted, ignored the simple things. Had he been so blinded by his own selfish desires that he had not taken the time to count his many blessings. Too close to life to see where real happiness lay?

In the drowsy in-between period before waking up, Tom wished he had savored the doing a little more, and the getting it done a little less. He wished he had openly acknowledged the gratitude due his family, his wife, told them he loved them more and shared more tears.

Then, rising from the shadows of things that had been, he saw himself as a child sitting beside his grand in the wood. An innocent boy, with his head on Grand's shoulder, their arms intertwined, he watched the sunrise with all the hopes and dreams for a future.

In his mind's eye Grand turned, and from the shadowy tangles of his mind came Grand's calm, comforting voice. "Don't you give up, Tommy. Don't you quit. You put your shoulders to the wheel and push through this thing. Don't lose heart or your faith in mankind." The reassuring warmth of Grand's words flooded over him. "With your hand in the hand of the Almighty to soften the pain, you keep on persevering one day at a time. I promise you, there is more happiness in your future."

Then, as quickly he had come, Grand was gone.

"You have to get up now, Tom!" came a warning voice dragging him from sleep. "Get up—get up now!"

Tom's eyes flew open.

The thundering sound of horses' hooves and a clattering wagon coming down the roadway drove his heart into his throat and drug his mind into full alert.

They're coming for me!

He grabbed the rifle and, laying camouflaged on the ramparts beside the roadway, hidden by the underbrush, he focused all his attention on the oncoming wagon. He could not yet make out the faces in the wagon, nor the trailing riders, rifles held high, ready for battle. Tom was determined to kill his pursuers without remorse, picking them off one by one, first with his rifle, then silently with his bow that lay beside him. With a little luck they wouldn't even see death coming, but it would surely come. He would take as many with him as he could.

"I will show you the same mercy you showed my family," Tom whispered. There was an unwavering sense of purpose in his madness. The hot crimson anger that had rushed to fill his heart had cooled into a hardened, calculated resolve as he leveled his rifle. "These men will pay for their murderous acts."

At deadly range now, Tom held the driver of the wagon in his sights. "Just a little closer."

He pulled back on the hammer one last click, readied the rifle to pull the trigger.

"Aim sure, shoot sure." Slowly he let his breath drain out, taking deadly aim on the driver. "You and I will be together in the fires of hell tonight."

He began to squeeze the trigger as the driver's eyes came into focus, but then paused in a moment of doubt before issuing the final death sentence. He blinked twice and pulled his head up from the rifle sights. The driver wasn't one of Boo Black's men. Tom's mouth went dry.

"Joey?"

Once, when Tom had been testing a new explosive mix in the mine, the explosion was so violent it slammed him against the wall. He hit the wall so hard, it knocked every wisp of air from his lungs. He was lying on the floor, face up, struggling to breathe, unable to speak. That's exactly how he felt when he realized he had almost shot his brother Joey dead.

His head was spinning, his hands shaking badly as he dropped the rifle to the ground. Still hidden in the underbrush, he stared speechless at the passing wagon. Then he saw Martha sitting beside his withered papa. It was that vision and the familiar whinny of Sadie and ole Blackie pulling the wagon that brought him to his senses.

"Mam?" A strangled cry went forth from his raspy throat.

Joey turned with a start, his rifle pointing through the misty morning in Tom's direction.

"Quiet, children," Andrew directed, his voice cracking in fear. "Get down. Don't say a word, whatever comes."

All waited in silence.

"Joey," Tom called out.

"Tommy?"

The wagon stopped.

"Tommy?"

Tom's muscles began to move again. He stood up in clear sight atop the brush-covered berm beside the roadway. Still dazed, he continued to stare at his family in confusion.

Joey uncocked the rifle and let out a long, shaky breath.

Tom's eyes settled on his papa, feeble, but still lean from all his years of shoveling coal. He wore his old dirty black slouch hat. His blue work shirt stretched over aging shoulders; his farmer jeans were held up by suspenders and a wide belt with his prized, shiny brass buckle. A wide smile eased across Papa's face as he looked back at Tom standing there.

Martha stood up beside Papa in her loosely fitting house dress extending down to her ankles, its paisley pattern long since faded into a pale grey. Her hair, threaded with silver, was gathered in a knot at the back of her head, and her scarred hands and arms were bare to the elbow. Calm, proud, and steadfast, she reflected a lifetime of service, dependability, and an unyielding determination to protect her family. Her intelligent amber eyes brightened at seeing Thomas there in the foggy mist. He thought them both so beautiful.

Martha brought her hand up to her mouth to hold in her emotion. "Oh, my boy! I had a bad feeling that I wasn't going to see you again, but here you are!" Her face was full of wonder. "Thanks be to God!"

"Mamma! Papa!" Tom's wits returning, now he saw Edie, Andrew, and their little ones, all sitting next to—Annie. "Annie! Oh, Annie," he whispered, tears rolling down his cheeks as he walked toward them. "I'm so sorry."

His hands trembled as he wiped at the tears in his eyes, crinkled at the edges in barefaced joy. Tension drained from deep within his soul. The bitter cup had passed. The dawn had broken, and all shadows of destruction fled as quickly as they'd come. His family was here before him, breathtakingly perfect in every way.

"My beautiful family!" Tom cried out in hysterical laughter, waving his arms wildly in the air. "Oh, dear God in heaven . . . thank you!"

He didn't care if he looked ridiculous scrambling through the brush and stumbling his way down the slope to the road. Still bewildered by their presence, but beside himself in unspeakable giddiness, he rushed to them, tears rolling down his cheeks. He was intent upon showering them all in unabashed hugs and kisses.

Tom grabbed the noses of the two horses. "Thank you! You big, beautiful ole mares. You shoulda been hauled off to the glue factory years ago, but thank God you were here to save my family. God bless you both!" He kissed them each soundly.

Then he came to Martha, tears rolling down his cheeks. He gave her an impassioned hug, holding on so tightly she could hardly breathe.

Tom turned to Papa, clasping his hand with both of his. He swept up the little ones and gave each a big wet kiss as they called out, "It's Thomas, Mam, it's Uncle Thomas! Do you see?"

There were hugs, kisses, and claps on the back all around for Joey, Edie, and Andrew.

Finally, he came to Annie. It became embarrassingly plain how inadequate language can be. She reached out for him, and he took her hands in his. "You have never looked so ravishing," he murmured hoarsely with tears in his eyes. "You are the most beautiful thing in all the world." He pulled her close. There were no more words. He kissed her long and hard, like someone drinking cool water after being desperately thirsty for days.

"Oh, Thomas," she held him tight, like she would never let him go, but could say no more through her tears.

"I thought you were dead," he whispered into her ear. "I found your bracelet in the ashes . . . Oh, Annie!"

"I fell and lost it in the fire." She could hardly contain her emotion as she held onto her husband, stuttering out her story. "We didn't see you . . . you didn't come . . . We had to leave. We broke an axle and had to fix it in the village back there . . . You must have passed us by . . . Oh, Thomas, I was so afraid for you."

"I saw the lock, the shuttered windows. Everything was in ashes. I was sure . . ."

"It was Papa," she interrupted, emotion flooding every part of her. "It was your papa who saved us all."

"Papa? But he can barely walk. How could he . . . ?"

Annie's words cascaded out. "There was no way out of the burning house. At the end, we all huddled together in the cellar. It was awful. The children were crying. No one could breathe. I felt like I was going to melt. I couldn't stop thinking of you and yet another tragedy you'd have to face. We all knew it was over for us, but no one could say it out loud."

"Oh, Annie!" Tom's dark eyes were wide with wonder.

"Martha was praying for a miracle when your papa burst into the cellar from outside. The oak door that had been barred shut. He'd been in the barn putting the final touches on the wagon, checking the straps and harnesses for the trip to Liverpool in the morning. When Boo threw open the barn doors, Papa hid in the hay 'til they were gone, and it was too dark to be seen."

"I don't think I would have been near so clever."

"All of us were suffocating from the smoke." Annie trembled as she recalled the moment. "We were confused when Papa J came through that cellar door, but the certainty of his words helped calm our panic. We just did as he said, ran out the door and escaped to the barn. Boo Black and all his thugs were either gone or so transfixed by the fire, they didn't even see us. We hid in the barn, then when we were sure the coast was clear, we left in the wagon as quickly as we could." Annie paused from the rush of the story to regain her composure. "Oh, Thomas. Your family lost everything."

"No—Papa saved all that matters." Tom hugged her tight. "Nothing else is important." Looking directly into Annie's wide-open eyes, he spoke solemnly, "I was so afraid I'd lost you!" Holding onto her shoulders, he held her gaze. "I love you, Annie. For all that you are."

Chapter 36

≽ · ≼

THAT EVENING, THE FAMILY SAT around the fire in the cold night air, not far from the outskirts of Liverpool. They stared quietly into the flickering flames that cast dancing shadows on the nearby trees. Only the crackle and sputter of the fire broke the silence.

Tom, his knees tucked up under his chin, arms folded over his legs, sat close to the scarlet embers to stave off the winter cold. Annie wrapped herself tightly into her husband, her arm threaded into his, her head leaning on his shoulder, careful to avoid his bandaged ear.

"I'm going back in the morning," Tom said, already hardened in his resolve.

"What?" Joey blurted out. "Going back where? You're leaving for America tomorrow." The others looked at Tom puzzled, awaiting his answer.

"I have unfinished business in Hoyle Mill." Tom was determined. "This is not the time to bend with the wind, but to fight it."

"They'll kill you," uttered Joey. A gasp went up from the family. Martha gave Joey a stern look, nodding toward Annie.

Annie took in a slow, shaky breath. She folded her arms to stop her hands from trembling and looked down at her feet.

"I'm sorry," Joey apologized.

"I've spoken to Annie and Mam about selling our tickets in Liverpool. That money should hold all of you over until you can find work. I hear there's plenty of work in that port town."

"Oh, Tommy," Edie interjected.

Tom ignored the objections. "I know some of you might be concerned that these tickets are not ours to sell. But our ox is in the mire, and it's easier to ask for forgiveness after. Besides, I suspect Bishop Walker can help in that regard."

"Are you sure, Thomas?" asked Andrew, his brows knitted together in concern. "All of Boo's men will be out for you and Lord Fitzwilliam . . ."

"Maybe, but what choice do I have?" Tom interjected, shaking his head. "Annie and I can't leave the rest of our family here with no money, no place to live, constantly looking over your shoulders for Lord Fitz's gang of assassins. Besides, they've done their worst. I think it best to publicly go on the offensive, fight back, and finish what needs to be done. If not me, then who will see this thing through?"

"I'll come with you then," Joey offered with conviction.

"No. You need to stay here to help the rest of the family. It'll take all of you working together to make this happen, to keep everyone safe and healthy. Thanks to Tutor, I have enough evidence to condemn them all. And with the help of Bishop Walker, hopefully we can use the publicity generated from the inquest to get to our new prime minister. He seems sympathetic to our cause and hopefully has enough influence to make use of our story and the many others in this valise. We can make a difference. Your task is to survive until I return. That won't be easy either. If you can make enough money over the next few months, we can pay back Mr. Stone and catch a brig with him to New York."

Stunned, the group let Tom's words lay quiet for a long, breathless moment.

Martha was the first to break the silence. "As much as I fear for my son, my daughter-in-law, and their new baby, I know deep in my soul, Thomas is doing the right thing." She looked at her son with admiration, this man that her little boy had become. Old memories buried deep in her heart of the night she left Woodhouse flooded her mind. She'd been a young girl in service to the powerful Fitzwilliam family those many years ago, when she was caught foolishly sneaking into the library late one night. The walls of the magnificent library were blanketed by a tumult of wonderful books that had drawn her helplessly in.

He had come at her drunken, with lust in his eyes. When she turned him away, he hit her with such force, it had loosened her teeth. He slammed her to the floor, knocking all the breath out of her, leaving her with no air in her lungs to scream as he forced himself upon her. With all her strength she had tried to fight him off, but it wasn't enough . . .

"Are you okay, Mamma?" Edie interrupted her thoughts.

"I'm fine—just fine." She collected herself and sat up straight. "You know the most damnable thing they did to me when I lived at Woodhouse Mansion?"

"What, Mam?"

"They taught me to read, to write, and to speak properly. And I shared that infinitely valuable blessing with all of you. We will all write wrenching letters to the newspapers. Tom will speak of atrocities to the villagers, and with Bishop Walker, to the politicians," Martha's strong, determined voice rung out into the night. "Yes, my son, you're right. You must give 'em hell. Do it Thomas! Do it! Make those damnable lords and ladies pay for the heartache they've wreaked. For

God's sake, don't worry about us. Do it, so we can all go to the promised land with our heads held high." Her eyes were wide with passion, her breath coming fast.

All looked at Martha in awe.

"I love you," Tom said, hugging her.

He turned and looked toward the others for confirmation. "I'm not leaving anyone behind. When this thing is done, we're all going to America. Together as a family."

Papa glanced over his shoulder at Martha sitting beside him. There was admiration in his eyes. "We'll get along just fine here, Tommy." He placed a comforting hand of support on Martha's shoulder. "Yer mam and I ain't dead yet. We got a bit of piss and vinegar in us still," he coughed out. "Don't ya worry 'bout this family."

?·?

In the days that followed, Tom learned Boo Black had not died from his injuries, though they were severe enough that he was past harm now. He would never again abuse miners or their defenseless families. Still Tom, emboldened by a heart filled with cold, hard resolve, was determined to see justice done. To make an example of Candyman Boo Black. Maybe, just maybe, there would never be a monster like him again. In pursuit of that quest, he joined with Bishop Walker.

With the stage set by the highly publicized inquest, they courageously took their case straight to the coal miner families in the village. Tom told his story of injustice in a voice rife with emotion. He looked upon impotent faces filled with rage and disgust, but disillusioned by the despairing frustration they were powerless to do anything about it.

In a mother's eyes, Tom saw his own mam's agony of heart. He saw in another heartbreak, discouragement, and a lost hope for her children. His heart ached when he saw his papa's broken spirit in the desolate eyes of a crippled miner struggling to stand, yet intent upon listening to every word. In the shadow of things that might have been, he saw himself as a beleaguered young boy in the slant of a young miner's shoulders, carrying the weight of shame for not being able to feed his family. His heart broke, jumping into his throat for protection. His somber eyes were shining when he saw the pain in the eyes of fathers and mothers who had been beaten down. These were men and women who had worked hard, endured much degradation, raising helpless children who too were being abused by their masters. Men and women left wanting at the hands of an aristocratic class who seemed bent on destroying everything they touched with their over-reaching, wrenching, grasping greed, unleashing company enforcers to devastate, shatter, and send families into mourning.

Both Tom and Bishop Walker spoke with passion of well-reasoned justice, of atonement, of change. And somewhere along the way in the heat of passion, Tom began to glimpse a glimmer of hope shining in the eyes of these men and women as Bishop Walker laid out the path in their redemptive quest, the path to make things right in their lives, in their families and communities. Tom saw in his neighbors, in his brothers of the mine, the pride of recognition, a badge of courage in their tattered clothing and hungry faces. He watched newly emboldened families stand together in resolve, declaring, "We ain't gonna take it anymore!"

"You must stand up for yourselves," Bishop Walker intoned. "Use your collective might to say, 'No more, we've had enough!'"

"Together we can make a difference," Tom told them. "Together we can make change happen." He saw strength in those who had watched their sons and daughters wither away under the oppressive, horrific acts by Boo Black and men like him in every coal miner's shantytown in South Yorkshire.

"Who will join with me?" Thomas called out, his heart soaring at the sight of the gathered crowds incited to fever pitch.

Men and women had hope in the eyes where none had been before.

"I will," came the cry of mothers, fathers, wives, widows, and grown children of those brothers in the mine lost or brutalized, with their families left behind to starve.

They rose their fists of support in anger, calling for revenge on those who had taken advantage of them, beaten them down, abused and raped their women, starved their children, and murdered their men.

"They've misused us long enough," the angry chant rang from those who had been forgotten.

Answered by their angrier wives and mothers: "And we ain't gonna take it no more."

Thomas hammered home the message. "We can make a difference if we stand and fight together. If we but turn the whole of our collective might against men like these."

Both Tom and Bishop Walker, working together, garnered a following of those who would not be pushed aside by men of power or the animals who brutally pounded out company demands on the innocent. When these miners had stood alone, they had been unable to fend for themselves. But together, they could withstand the onslaught of company brutality and landlord demands to buckle under to the customary rules of industrial slavery.

These men and women gathered with them as they walked down the village streets, in an ever-growing movement to share their own harrowing experiences of death and destruction at the hands of company henchmen. Candyman Boo Black, despite being a shell of the man he once was, became the symbol of tyranny. Soon, every man, woman, and child in South Yorkshire knew the name of the man who had burned down Thomas Wright's family home, and but for the grace of God, would have murdered his entire family. Their own anger, having simmered for decades, now reached a boiling point, and the dragon was unleashed. The power of rage rose to a frenzy in the many company-owned shantytowns and villages, pushing forward their cause, with hundreds of seething men and women joining in. Roaring in emotional outcry for justice and change, they took their cause all the way to Westminster, the pulsing heart of England, where their stories splashed across the pages of the London press.

Bishop Walker, who had attended Oxford as a young man, took advantage of knowing friends in high places. Through his efforts and others, like Lord Ashley, buoyed by the support of Prime Minister Disraeli, they would push for new laws exacting consequences for the irresponsible actions of mine operators and landlords.

In the end, no one defended Boo Black in the court of public opinion nor in the court of law, least of all the aristocracy for which he worked his evil deeds. The company and its aristocratic patrons were stunned and enfeebled by the tumultuous scandal. They took the path of least resistance, condemning Boo Black for his part in it—and throwing him to the angry mutineers. Boo would be convicted not only of the attempted murder of Thomas's family, but the successful murder of other miners and the rape and ruthless beating of family members who complained.

⋟ · ⋞

It had taken many months, but now Tom stood among the angry crowd, cloaked in a concealing coat and hat, his face stern as hammered iron. With a rolling, queasy stomach, he looked on at Boo Black as he shrank before the Newgate Prison gallows. Tom watched Boo stumble along the pathway heading to the hangman's scaffold, hands tied behind his back. As Boo passed by, some in the crowd cheered. Some muttered curses. Here and there were echoed gasps of outrage. Cries of disgust toward this evil man were issued forth from the furious mob. Someone in the front of the gathered onlookers spit in his face. All seemed pleased to see the last of Candyman Boo Black.

"History promises this for certain," the newspapers read, "we get what we deserve."

Boo Black would make a bad example of how one should approach death by hanging. The three burly guards were forced to drag him staggering over the final yards to the gallows. He kicked and screamed, his wild eye bulging with terror, his mouth hanging open from his red puffy face, drool dripping off his raggedy chin. There was a widening stain on his dirty brown prison cloak where he wet himself. He smelled of feces. Wild-eyed and crying like the baby he had once sadistically torn from a mother's arms. He tried desperately to jerk himself free from his tormentors.

Frustrated with his struggling, the guards pulled a gunnysack over his head, then dragged him up the stairs to the platform like a blind pig going to slaughter. They needed to hold him down on the platform, remove the gunnysack, place the noose over his head, then lift him up again before dropping him through the open hole in the platform to hang.

The rope cinched so tight around his neck from his struggling, he died in the most horrifying way. As he hung from the gallows, he suffocated to death like those in the afterdamp, silently jerking and twisting, opening and closing his mouth like a fish out of water.

The mob grew silent as his death ritual dragged on.

Tom stared aghast at the horrific scene, deriving no pleasure from it. With an acrid taste in his mouth, he turned away in long strides and left the gathered mob before the appalling conclusion to this pathetic man's degenerate life.

Leaving the yard, he heard a young boy's anguished cry, "Please, Papa, make this end."

As he walked out the gate, Tom had one thought on his mind: getting back to the comforting fold of family.

⋟ • ⋞

Dear Mrs. Annie Wright,

I sent your husband home to you today. I wanted to thank you for loaning him to us over these past several months. Thanks mostly to his efforts, we have sent Candyman Boo Black off to meet his Maker. I'm sure that will not be a pleasant reunion, maybe even more distressing than his hanging from the gallows. He has a lot to answer for. And in the months ahead, I suspect the aristocratic industrial slavers who own the coal mines will have many sleepless nights as well, haunted by the stories told in front of Parliament. Stories of their oppression of those who were supposed to be under their watchful care. Stories that have appalled a sympathetic Queen Victoria and caring men and women all across England.

I wouldn't want to be in the shoes of these arrogant lords and ladies when they are received at the foot of the great Almighty.

Contrast that with your husband. When he returns home to you and the rest of his loving family, he will lie beside you at night, his conscious clear, knowing Candyman Boo Black is no more, and the plight of the coal miner in England is better off because of his efforts.

God bless you all for your sacrifices. And I pray you and Thomas have a happy life.

Bishop Walker

Chapter 37

⤜ • ⤛

A FTER ALMOST SEVEN MONTHS TOM returned to his family in Liverpool. Living in a small tenement off Main Street in the bustling port town, they had worked, collectively saving their money over the winter for the passage to America. Martha especially had done well for herself, finding a home writing for the Liverpool *Mercury News*. The family had saved almost enough to pay for the venture across the Atlantic and were in the midst of planning the trip with Tom's future employer, The Comstock Mining Company. To make up the financial shortfall for the voyage and trek west for the entire family, the brigantine *Colorado* had consented to take on Tom and Joey as a part of the crew.

Tom walked into the tiny tenement unannounced. All in the family were there preparing for their evening meal. He hadn't realized how much he'd missed them until he saw them all laughing and working on dinner together.

Annie spun around in surprise, her hand flying to her mouth. "Thomas!" Almost in unison the entire family turned and rushed him with hugs and kisses.

David A. Jacinto

Filled with elation at seeing his very pregnant wife, Tom threw his arms around her as the rest of the family looked on with smiles and giggles. "Oh, Annie, look at you—you beautiful thing."

"Humph! We may have to get a bigger boat," Annie chortled, wiping the tears from her elated eyes. "I missed you so." The weight of the past several months had been lifted. She could breathe again. "I can hardly believe it," she shuddered out, "here at last."

Tom put his arms around them all. "There will be no one chasing us anymore," he whispered. "We are free to start our new life, unshackled."

The heavy burden lifted, they laughed and cajoled each other and weren't quite sure what to do next but smile, smile, and smile some more.

With his arm draped over Annie's shoulder and both of hers around his waist, he whispered in her ear, "I don't suppose I deserve to be so happy."

Annie only smiled.

⤝ • ⤜

When the dust of reunion settled, Martha insisted that while the rest of the family headed off to America, she and their ailing papa would stay in Liverpool. She would continue to write for the *Mercury* and take care of Papa. "At least until you children are settled, and Papa feels better."

They each knew the time to see their aging parents again would never come. But despite the protests and arguments made to change their minds, both Martha and Joseph insisted they stay in their homeland. They were resolved to give what money they saved to their children and grandchildren for their journey west. In this unselfish

⤝ 376 ⤜

offering, Martha quietly made her final act of sacrifice for her children, taking on the burden of providing for their feeble papa, nursing him to the end, fully expecting to go on to rest in an unvisited grave.

≯ • ≺

On the day of their departure, the Liverpool docks were bursting with the hustle and bustle of travelers bound for destinations around the globe. More ships came and went from Liverpool than any other port in the Northern Hemisphere. And with that distinction came an astounding number of thieves, confidence men, pickpockets, and swindlers milling through the crowds of beleaguered travelers, looking for an opportunity to separate the trusting from their money.

When the Wrights' wagon lumbered to a stop, all in the family traveled down to the wharf, each in pursuit of fulfilling their assigned tasks. An exhausted Tom stepped down from the wagon, trying to yawn away the last few sleepless days of travel from his bones. Not thinking, he set his knapsack down on the dock. It contained all their valuables, including tickets for the brigantine *Colorado*.

No sooner had the knapsack hit the wharf then a runner, quick as a flash of lightning, snatched it up and raced off. Tom turned around just in time to see all their valuables vanish faster than he could ever hope to catch up.

"Stop that man!" he called out as the runner dodged through the crowd.

Unfortunately for the runner, he ran headlong into Annie Dale Wright buying feed for their horses. The thief didn't see the fifty-pound bag of oats in Annie's hands, or maybe he didn't comprehend what a weapon it could be in the hands of a young woman in her delicate condition, who was accustomed to lugging sides of beef for a living. With a roundhouse swing of the gunnysack, she knocked the bandit sprawling across the dock.

Joey, standing nearby, quickly rescued the knapsack. And the runner was off with no wages for his effort, nor compensation for his pain and suffering. His only keepsake from the run-in with Annie's right hook was a black eye, a headache, and a weakly fabricated story to save embarrassment with his ring of grifters.

"There's ice in those veins." Joey smiled at Annie with admiration. "You're a force to be reckoned with!"

"Finally, something good from all my totin' and luggin'." Annie blew out a sigh of relief.

"I'm afraid that poor fool believed she was of the weaker sex," Martha nodded toward Annie with a knowing smile.

For a brief moment, all delighted in their brush with financial ruin.

Tom stopped a longshoreman. "Excuse me, sir. Can you please point me in the direction of the brigantine *Colorado*?"

"That'd be her, young man." The longshoreman pointed Tom to a small, battered brigantine packet ship easing into her slip alongside the dock. "She be your *Colorado*."

"Oh, no!" Edie gasped when she saw the squat, ugly little brig where they would spend the next two fortnights. "It hardly looks seaworthy enough to sail us up the Thames, let alone . . ."

"Cross the Atlantic," Joey whispered. "Do you really think we're crossing in *that*?" All watched it being tied up to the dock. "It says *Colorado* on the keel."

"Oh, Mam, I'm so excited. I just can't wait," added Elizabeth, looking at the *Colorado*.

Edie rolled her eyes. "Sometimes I wonder if she's really ours, Andrew."

Long past her prime, this ninety-one by twenty-seven-foot, double-masted packet ship drew only a dozen feet when loaded. "It looks like the smallest vessel to cross since the *Mayflower*," Annie snickered. "Better load one family at a time—she might flip her over if we all step onto the gangplank at the same time. But then at least we'll drown together as a family."

"That's not funny, Annie," Edie chided halfheartedly.

"Right. At least Thomas won't be havin' no more worry 'bout findin' his little brother a wife," Papa chimed in.

Joey scoffed. "Are you kidding? The day I let Tom pick my wife is the day I head back to England. He's got no clue!"

Annie put her hands on her swelling stomach. "I'm insulted. I thought I was the best thing that ever happened to your brother?" Annie reproved with a smile. "You're not getting any Christmas presents from me."

"That was dumb luck, Annie. I still don't know how he talked you into marrying him."

"Well, I really put up a fight, didn't I?" She raised her ample brows.

With tearful hugs, bittersweet smiles, and goodbye kisses, Martha and Joseph reluctantly said their final farewells to their children and grandchildren, knowing they would grow up without her and Papa in their lives.

Pulling Annie aside, Martha laid a calloused hand against her cheek and looked solemnly into her sober eyes. "I've come to know you as a strong, competent woman, capable of great loving-kindness. You are the best thing that ever happened to Thomas."

"Thank you, Martha," Annie sighed in a thready voice. "I can't tell you how much it means to hear you say that."

"But you can't believe it, can you?" Martha sighed. "Because you are still comparing yourself to Lydia."

Annie blinked back tears. "She was so beautiful. Thomas's soul-mate. How could I ever . . . ?"

"And you believe what you see in the mirror makes all the differ-ence?" Martha interrupted. "I don't know much about that kind of beauty, but I know this—you are strong, have heart, and an inward beauty and determination to match. A very rare commodity. Thomas may not fully see it just yet, but he will. I know in my heart you will prove to be the completion of him."

Annie wiped at her welcoming eyes. "Thank you, Martha. You've become my mam more so than my real one ever was."

Martha paused for a moment. "You must promise me one thing. Edie is a wonderful mother who loves her children to a fault, but she will need your strength to survive this great adventure on which you are about to embark. Can you promise me that you'll be there for her?"

Annie couldn't speak. She reached for Martha's hand and worried that once she touched this woman she had learned to love, she might not be able to let go. She stared at her, memorizing every facet of her face, the face of the mother she loved but would most likely never again see. "I will."

"Thank you." Not wanting to burden Annie any further, Martha turned back to her children, her eyes moist. "Godspeed, my lovelies!" Tears ran down Martha's face, and she turned to Thomas. "Promise me you'll keep my grandchildren safe."

"I promise, Mam!" Tom's answer came in a thready voice.

Martha stared a long moment at her son. "I'm gonna miss you. I had every intention of watching you take the world by storm, my beautiful, determined boy."

Tom could barely speak. "No one will ever love me like you do, Mam."

"Don't count on it, Tommy. You've married a good one."

He smiled. "You are the only thing holding me here. I will miss you every day for the rest of my life."

She reached up and ran the back of her finger down his cheek, tears trickling down her own. "Find your way in America," Martha whispered. "Never give up until you've reached your dreams. But remember, family first. Be kind to Annie—she loves you so. She will be a wonderful wife and mother. And don't let your little ones forget Papa and me."

"Yes, Mam."

≯ · ≺

Having said their final goodbyes, Tom, Annie, and the rest of the family walked up the plank onto the *Colorado*. They would be aboard an American ship now, with a freewheeling American captain, who, as he put it, "didn't cotton much to the Brits."

Annie stood with Thomas at the railing on the aft quarter deck of the little brig, their arms intertwined. She laid her head against his shoulder. The rest of the family accompanying them on the voyage were also huddled nearby. They gazed toward the Liverpool docks and all that they had ever known fading into the distance as the *Colorado* pulled away from the harbor. This was a ship filled with human souls fleeing the tyranny of an old-world Europe, seeking a new life in the new world, bound together by a lust for freedom. Not only for themselves, but for their children and their children's children—and not just English children, but the Scots, Irish, French, Dutch, Swedes, Germans, and Jews from every port in Europe. Tom and his extended family of ten souls shared with these unlikely companions the grief-stricken, wrenching loss of their homeland and family left behind and a reticent, uncertain hope for the future that lay ahead. In this uncommon company, with common dreams, they would face the daunting voyage together, through howling ocean storms, over tempestuous seas, and driving wind and rain.

"Freedom is a powerful force . . . for all men," Tom offered wistfully. "A chance at a better life, to rise above our shackled existence." He pondered the thought. "Unleashing the great sweep of human desire. That's what will make America the greatest country ever. What was it Abraham Lincoln said, Annie? 'Folks are usually about as happy as . . . ?'"

"'. . . as happy as they make their minds up to be,'" Annie finished Lincoln's sentence.

"Are you happy, Annie? I've taken you away from Emma and your family, and all you've ever known."

"You are my family now." She put her free hand on the swell of her belly. "So, unless you plan on throwing us overboard before we make landfall in America, I feel surprisingly good about our uncommon affair." With a gentle smile she tightened her comforting hold on his arm.

Tom thought of those who had brought him here. Grand, Tigger, and Lydia, who in death had opened a window to his soul. And his courageous mam, who helped him to see the light, and the possibility of pursuing his dreams. And of course, Annie, who pushed him into discovering the most priceless things in life are not really things at all—gratitude, hope, love, and family.

"I think courage could be the rarest of all virtues, don't you, Annie?"

"Some say it's the cornerstone of change."

He thought for a moment. "Not just the courage to beat back our enemies. But the courage to put aside fear and do the right thing. It's where character is forged, isn't it?"

"And the ties that bind family together . . . strengthened."

Tom smiled, then turned and whispered in her ear, "You saved my life, you know." His breath caught. "You helped me find my way out of the darkness."

Looking out to sea, she smoothed the apron over the belly that held their child, but said nothing more. She couldn't.

Acknowledgments

THE TELLING OF THIS STORY would not have been possible without the work of special people who had a great influence on me:

My great-aunt Norma Jean Wright-Trietsch, who spent the first half of the twentieth century in a Herculean effort to meticulously comb through and organize genealogical records, old letters, journal entries, news clippings, and other documents, and who interviewed countless relatives with connections to the nineteenth-century lives of the Wright family.

The National Union of Mine Workers, Yorkshire, England, and the Daughters of the Mineworkers and their willingness to provide information on mining and the Oaks Mine and Silkstone Mine disasters.

Michela Miller Dickson, a delightful young woman who spent countless hours going through this manuscript with me and engaging in long conversations to make sure I got it right.

Michael Levin, a brilliant writer himself and winner of multiple awards, who coached me through this process. He taught me much about writing good historical fiction and how to create a storyline that does justice to this absolutely fascinating story.

And of course, Christina Boys, for her highly skilled and insightful editing, and Jonathan Merkh, Justin Batt, Jennifer Gingerich, and the rest of the Forefront and Simon & Schuster team.

Acknowledgments

I couldn't conclude my acknowledgements without mentioning my family: Thomas and Annie Wright. Their parents, the invincible Martha Wright and her husband Joseph. Joseph and Edith Wright. My daughter Rachel for slogging through the first of the manuscript and offering valuable suggestions. My wife for reading the last draft and adding her suggested changes.

Author's Note

⋟ · ⋞

THOMAS CAME OF AGE IN a time of struggle for the working class during the chaotic Industrial Revolution that raged across England. At the time of this novel, England ruled Europe, and Europe ruled the world, while America (covered in Book II of this series, *Where Eagles Fly Free*) was still a tenuous upstart on the world stage. Some would call this a story of ordinary people, but it is often ordinary people doing extraordinary things that change the world for the better, then go quietly to uneulogized graves. These are the kind of ordinary people who, in the nineteenth century, immigrated to America to make it the greatest country the world had ever known.

Thomas, Annie, Lydia, Martha, and the rest of the Wright family are reflective of these ordinary people—real people shown in the family tree exhibit included herein. The dates, places, events, and many of the personal incidents are generally accurate, although timing was sometimes altered for brevity. Much of what I know of Thomas, Annie, and the rest of the Wright family was taken from a surprising number of genealogical records, testimonials, letters, journal entries, and documents passed down to posterity. Those two large trunks referenced in the introduction, including Martha's ships locker dating from 1810, are now in my attic with Annie's simple bracelet, the *Colorado*'s ship log, Annie and Thomas's marriage license, and many other things. A portion of the family estate on Chalk Creek, Spring Hollow (referenced in Book II), has been preserved as a historical monument.

As with most works of historical fiction, the most outlandish events are often the true ones. Thomas and his papa, living in a small miner's shanty with his family, did each start working in the coal mine at age seven under horrific conditions, his grand at age five. Thomas, his father, and grand were all involved in serious accidents as children working in the mines. They carried scars for the remainder of their lives. I can only assume both physical and emotional scars helped shape and mold Thomas's view of life.

The Wright family and wider community in which they lived were governed by the law and rules of conduct laid down by the fantastically wealthy Lord Fitzwilliam, a mine operator, landlord, and member of Parliament, who controlled 85,000 acres of land and lived on what could arguably be called the most magnificent estate in all of Europe. Situated on the 14,000-acre Woodhouse Estate, Lord Fitzwilliam's 255,000-square-foot, 310-room mansion was so enormous that Darcy's mansion from Jane Austen's *Pride and Prejudice* would be a mere guesthouse by comparison. Walking from room to room, it would take one of the more than four hundred and fifty servants—some records suggest up to one thousand staff at certain times in its history—an entire twelve-hour workday just to walk from one end of the mansion to the other. It is now the Wentworth Woodhouse Museum.

George did become an American journalist until he died of unknown causes as a young man. Thomas and Lydia were married only a week before she passed away, and Thomas disappeared for a time before he married a much younger Annie Dale a year later without an engagement. The death of Thomas and Annie's first child was difficult on both parents. The *Colorado's* voyage across the Atlantic included all ten of the family members referenced, but the descriptions of the ship and voyage were taken in part from Thomas's brother George's earlier journal on a brigantine sailing ship to New York harbor.

On July 4, 1838, twenty-six children died in the Silkstone tragedy, ages seven to fifteen. Eleven of them were girls. All were buried in mass graves for reasons of expense. Although the circumstances of their deaths were modified for purposes of the story, the twenty-six children included Tigger (John) Gothard, age nine; Fidget (James) Turton, age nine; Katie Garnett, age ten; Abe (Abraham) Wright, age eight; and Zac (Isaac) Wright, age twelve. Mary Wright was left with three little girls and no means of support after the deaths of her husband to firedamp in an earlier explosion and the 1838 deaths of her only two sons, Abraham and Isaac. Not long after her losses, Mary and her three little girls were kicked to the streets to starve by company thugs, along with other families in Silkstone. In all, 370 British colliers were killed in 1838 alone, and eighty-six were children. Both the Silkstone and Oaks mining disasters were deemed accidental acts of God. No one was held responsible for the incomprehensible lack of safety measures in these mines.

In the Oaks mining disaster on December 12–13, 1866, tragically 383 men and boys were killed. Many of the stories of the Oaks tragedy told here were taken from the actual records of the disaster. For example, Emma's Billy Abbott was rescued and died from his burns; Jenny Lewton did lose her husband and seven of her eight children in the disaster, and like all the rest, was left with no company compensation for her losses; a poorly supplied onsite triage was set up by young Doctor Blackburn; and except for timing and the inclusion of Tom, the rescue of Samuel Brown happened as described. The facts surrounding Will Wright, his family, his unconscionable incarceration, and his death in the Oaks explosion are true.

Of course, Thomas and Annie's involvement in the historically significant events depicted in these two books are not entirely accurate. These are novels, after all. And as is the nature of most historical fiction worthy of the name, this tale has been told with the requisite embellishments where there are no eyewitnesses alive to spoil the story. Their involvement in the major historical events of their time and place were often fictionalized. Where there are neither supporting nor detracting documents to suggest otherwise, I've chosen to allow these real-life characters to be major participants in these historically significant events for the purposes of the novel. For example, because of Tom's connection with the Silkstone & Elsecar Coalowners Company and the Oaks Mine, it is likely he would have been involved in the rescue attempt that killed engineer Parkin Jeffcock and twenty-five other rescuers, although there is no documentation of Thomas, nor most of the other more than a hundred rescuers who lived through it.

Bishop Walker is a composite character of several people, including Lord Ashton, a local bishop, other clergy, and Mr. Channell at the inquest. Much of the dialogue at the Hoyle Mill inquest was taken directly from historical records, although Thomas's testimony is entirely fictional. Tutor Turton is a fictional character, but the actual company pit manager who was sympathetic to the miners was also forbidden from testifying at the inquest. The burning of the Wright house was purely fictional, fabricated from other events for purposes of the novel. Boo Black is also a fictional character, depicting actions not untypical of thugs for hire at the bidding of landlords carrying out operation companies' deplorable abuse of coal miners and their families at the time.

After the Silkstone disaster, Parliament, in a hard-fought legislative process, passed the 1842 Mines and Collieries Act, prohibiting young girls from working in coal mines and boys younger than ten, though entire families often continued to work together down in the pit. Much of the public outcry was driven by mothers of these children who, like Martha, encouraged Lord Ashton to pursue the legislation. Charles Dickens wrote an impassioned letter to Parliament in support of the bill the day before it was heard and it is said the plight of these child miners was the impetus for his "A Christmas Carol," written the following year in 1843. That same year, the Silkstone tragedy and subsequent bill inspired Elizabeth Browning's poem "The Cry of the Children," in which she wrote, "A child's sob ringing out in the silence is a curse far more powerful than even a strong man's rage."

After the Oaks Mine "Christmas Disaster" in December 12–13, 1866, it took Parliament six years to pass the Coal Mines Regulation Act of 1872, because of manipulative delays championed by Lord Fitzwilliam and other Members of Parliament with coal mining interests. This new law restricted working age to boys twelve and older, required strict third-party inspection for safety measures, and ascribed limited liability to the mine operators and landlords in future disasters. Both pieces of legislation would save thousands of lives.

The cause of Lydia's death after only one week of marriage is unknown, but more than 20 percent of all girls and young women who worked in match factories died of phosphorous poisoning between 1856 and 1880. A later "Match Girls" strike led to safety laws protecting women and children in this industry and eliminating the fourteen-hour workday, where Match Girls like Lydia lived and died.

Shortly after the Wright family left England for America, Prime Minister William Gladstone testified in a speech to Parliament, "The Constitution of the United States is the most wonderful work ever struck off by the brain and purpose of man." Echoed the celebrated prime minister, "America is a state of mind, a blessed land where all the hopes and dreams of life are possible. Where liberation of the human spirit allows freedom of thought; the freedom to dream; the freedom to reach out and taste the excitement of life. The opportunity to build and fashion a life of your own choosing; it is all there for the taking in America; opportunities that most peoples around the world dare not even dream about."

Bibliography

Addey, John. *A Coal and Iron Community in the Industrial Revolution.* Harlow: Longman, 1969.

Bartoletti, Susan Campbell. *Growing Up in Coal Country.* Boston: Houghton Mifflin Company, 1996.

Britannica. "Benjamin Disraeli: Quotes." Accessed December 2021. https://www.britannica.com/quotes/Benjamin-Disraeli.

Chamberlin, E. R. *The Awakening Giant: Britain in the Industrial Revolution.* London: B.T. Batsford Ltd., 1976.

Coal: British Mining in Art, 1680–1980: an exhibition. Brighton: Arts Council of England, 1982.

Darlow, Paul. *The Oaks Disaster 1866: A Living History.* Self-published, undated.

Dickinson, Joseph. "Oaks Mines Inspector Report." 1866.

Disraeli, Benjamin. 1983. *Sybil, or The Two Nations.* London: The Folio Society. First published in 1845.

Engels, Friedrich. 2009. *The Condition of the Working Class in England.* Oxford University Press. First published 1845.

Gallop, Alan. *Children of the Dark: Life and Death Underground in Victoria's England*. Stroud: Sutton Publishing, 2003.

George, M. Dorothy. *England in Transition*. Milton Park: George Routledge & Sons, Ltd, 1931.

Gibbons, Boyd. "The Itch to Move West." *National Geographic*, August 1986.

Gladstone, William Ewart. "Published Speeches to England's Parliament." 1868–1874.

Hamblin, Jacob. Journals and Letters to Jacob Hamblin. Manuscript copies prepared by Brigham Young University Library, 2004.

J.C. 2009. *The Complete Collier: or The Art of Sinking, Getting, and Working Coal Mines, Etc. As Is Now Used in the Northern Parts*. Whitefish, MT: Kessinger Publishing. First published in 1708.

Lake, Fiona and Preece, Rosemary. *Voices from the Dark: Women and Children in Yorkshire Coal Mines*. Wakefield: Yorkshire Mining Museum Trust, 1992.

Manchin, Frank. *The Yorkshire Miners: A History*. Barnsley: National Union of Mine Workers, 1958.

Meese III, Edwin. "The Meaning of the Constitution." The Heritage Foundation Political Process Report. Published September 16, 2009. https://www.heritage.org/political-process/report/the-meaning-the-constitution. [(1878 quote by Prime Minister William Gladstone). Political process report, heritage.org.]

Prince, Joseph F. *Silkstone: The History and Topography of the Parish of Silkstone in the County of York*. Penistone: J. H. Wood, The Don Press, 1922.

Roberts, Robert A. *Of Masters and Men: The Clarkes of Silkstone and their Colliers.* Sheffield: Barnsley Workers' Educational Association, 1981.

Sheffield Archives. "Records of the Clarke Family of Noblethorpe Hall, Silkstone." (NRA Ref: 6593).

Stowe, Harriet Beecher. *Little Foxes.* First published 1866.

Teasdale, G.H. *Silkstone Coal and Its Collieries.* Privately published, undated.

Unreferenced. "Colliery Guardian Article." Barnsley Chronicle. 15th December 1866.

Unreferenced article. Colliery Guardian. 15th December 1866.

Unreferenced. "Great Pit Disasters in Great Britain. 1700-1866." Sheffield chronical. 1866.

US Department of Labor. "Title 30—Mineral Resources." Mine Safety and Health Administration, Department of Labor. Last modified April 20, 2023. https://www.dol.gov/general/cfr/title_30.

Wikipedia. "Oaks Explosion." Last modified December 13, 2022. https://en.wikipedia.org/wiki/Oaks_explosion.

[Queen Victoria's quotes and contribution and Prime minister Benjamin Disraeli quotes on the Oaks Mine disaster 1866-67), (accessed 2021)]

Wikipedia. "Wentworth Woodhouse." Last modified April 5, 2023. https://en.wikipedia.org/wiki/Wentworth_Woodhouse.

Wikipedia. "Wentworth Castle." Last modified March 30, 2023. https://en.wikipedia.org/wiki/Wentworth_Castle.

Wikipedia. "Woodhouse, South Yorkshire." Last modified September 18, 2022. https://en.wikipedia.org/wiki/Woodhouse,_South _Yorkshire. [Washlands nature Reserve, 2020]

Wright-Trietsch, Norma Jean. *They Came from England: The Wrights, 1860–1972,* self-published, 1972.

About the Author

DAVID JACINTO WAS BORN INTO a family living on the wrong side of the tracks and has been a storyteller ever since. He was the first in his extended family to attend college, and as a student athlete at one of the most prestigious universities in the country, he received his degree in civil engineering. He went on to serve as a president of SM Engineering Company, held leadership roles in multiple national and international companies along the west coast of California, and was on the board of directors for a few more. He was also commandeered by the State of California on special assignment as chief engineer to help rescue California's three major utilities on the verge of bankruptcy during the highly publicized, $300 billion energy crisis in 2001.

David has had numerous speaking engagements over his successful career, frequently interjecting colorful, fascinating, and humorous stories of his life experiences. Some of these stories are drawn from his ill-spent youth, some from his many business successes, and some from family experiences. But all are delivered with the greatest respect for the opportunities America has afforded him and a thankfulness to those fallen leaves from the family tree of immigrants who made it all possible.

Despite his business successes, he supposes his greatest achievements have been to convince the fetching Anne Gray to become his wife, the good fortune to be a part of the lives of his four wonderful children, their wives and husband, and the blessing to be Papa J to thirteen near-perfect grandchildren.